# TOPICS IN DYNAMICS

## I: FLOWS

BY

EDWARD NELSON

PRINCETON UNIVERSITY PRESS

AND THE

UNIVERSITY OF TOKYO PRESS

PRINCETON, NEW JERSEY

1969

L.C. Card: 79-108265

S.B.N.: 691-08080-1

A.M.S. 1968: 3404, 4615

Published in Japan exclusively
by the University of Tokyo Press;
in other parts of the world by
Princeton University Press

Printed in the United States of America

# I. FLOWS



These are the lecture notes for the first term of a course on differential equations, given in Fine Hall the autumn of 1968.

It is a pleasure again to thank Miss Elizabeth Epstein for her typing.

# I. FLOWS

In classical mechanics the state of a physical system is represented by a point in a differentiable manifold M and the dynamical variables by real functions on M. In quantum mechanics the states are given by rays in a Hilbert space $\mathcal{H}$ and the dynamical variables by self-adjoint operators on $\mathcal{H}$. In both cases motion is represented by a flow; that is, a one-parameter group of automorphisms of the underlying structure (diffeomorphisms or unitary operators).

The infinitesimal description of motion is in the classical case by means of a vector field and in quantum mechanics by means of a self-adjoint operator One of the central problems of dynamics is the integration of the equations of motion to obtain the flow, given the infinitesimal description of the flow.

## 1. Differential calculus

In recent years there has been an upsurge of interest in infinite dimensional manifolds. The theory has had important applications to Morse theory, transversality theory, and in other areas. It might be thought that an infinite dimensional manifold with a smooth vector field on it is a suitable framework for discussing classical dynamical systems with infinitely many degrees of freedom. However, classical dynamical systems of infinitely many degrees of freedom are usually described in terms of partial differential operators, and partial differential operators cannot be formulated as everywhere-defined operators on a

Banach space. We will be concerned only with finite dimensional manifolds. Despite this, I will begin by discussing the general case. I do this for two reasons: because the theory is useful in other branches of mathematics and because the fundamental concepts are clearer in the general context.

Let $E$ be a real Banach space. That is, $E$ is a real vector space with a function $x \rightsquigarrow \|x\|$ mapping $E$ into the real numbers $\mathbb{R}$ such that $\|x\| \geq 0$, $\|x\| = 0$ only if $x = 0$, $\|ax\| = |a|\|x\|$, $\|x+y\| \leq \|x\| + \|y\|$, and $E$ is complete: if $\|x_n - x_m\| \longrightarrow 0$ there is an $x$ in $E$ with $\|x_n - x\| \longrightarrow 0$. For example, $E$ may be an s-dimensional Euclidean space $\mathbb{R}^s$ in the norm $\|x\| = (x_1^2 + \ldots + x_s^2)^{\frac{1}{2}}$. If $F$ is another Banach space we denote by $L(E,F)$ the Banach space of all continuous linear mappings of $E$ into $F$ in the norm $\|A\| = \sup\{\|Ax\|: \|x\| \leq 1\}$. We abbreviate $L(E,E)$ by $L(E)$.

Let $U$ be an open subset of the Banach space $E$, and let $x$ be in $U$ A function $f: U \longrightarrow F$ (where $F$ is a Banach space) is said to be (Fréchet) <u>differentiable at</u> $x$ in case there is an element $Df(x)$ of $L(E,F)$ such that

$$f(x+y) = f(x) + Df(x)y + o(y),$$

where $o(y)$ is a function defined in a neighborhood of $0$ such that $\|o(y)\|/\|y\| \longrightarrow 0$ as $y \longrightarrow 0$ with $y \neq 0$. It is clear that $Df(x)$ is unique if it exists. It is called the (Fréchet) <u>derivative of</u> $f$ <u>at</u> $x$. The function $f: U \longrightarrow F$ is called <u>differentiable</u> in case it is differentiable at all points $x$ in $U$, and it is called $C^1$ in case it is differentiable and $x \rightsquigarrow Df(x)$ is continuous from $U$ to $L(E,F)$. If $f$ is $C^1$ then $Df$ is a function from $U$ into the Banach space $L(E,F)$, so it makes sense to ask whether $Df$ is $C^1$.

The function $f$ is said to be $C^2$ in case $f$ is $C^1$ and $Df$ is $C^1$ and, by recursion, $f$ is said to be $C^k$ in case $f$ is $C^1$ and $Df$ is $C^{k-1}$. (A trivially equivalent definition is that $f$ is $C^k$ in case $f$ is $C^{k-1}$ and $D^{k-1}f$ is $C^1$. Sometimes one definition and sometimes the other suggests the more convenient way to organize an induction proof to show that $f$ is $C^k$.) Similarly, we define $f$ to be $k$ <u>times differentiable</u> in case it is differentiable and $Df$ is $k-1$ times differentiable (or equivalently, in case it is $k-1$ times differentiable and $D^{k-1}f$ is differentiable). Notice that if $f$ is differentiable at $x$ it is continuous at $x$. Consequently a differentiable function is continuous, and a $k$ times differentiable function is $C^{k-1}$.

Let $E_1,\ldots,E_n$ be Banach spaces, and consider their Cartesian product $E_1 \times \ldots \times E_n$. It is possible to give this a Banach space structure by defining addition and scalar multiplication componentwise and giving an element the norm which is the sum of the norms of its components. This Banach space is denoted by $E_1 \oplus \ldots \oplus E_n$ and called the <u>direct sum</u> of the Banach spaces $E_1,\ldots,E_n$. Elements of it are denoted by $x_1 \oplus \ldots \oplus x_n$, where $x_i$ is in $E_i$. Frequently we wish to consider multilinear forms on $E_1 \times \ldots \times E_n$; that is, functions on $E_1 \times \ldots \times E_n$ which are linear in each variable separately. If $F$ is also a Banach space, we let $L(E_1 \times \ldots \times E_n, F)$ be the Banach space of all continuous multilinear forms on $E_1 \times \ldots \times E_n$ with values in $F$, with the norm

$$\|A\| = \sup\{\|A(y_1,\ldots,y_n)\| : \|y_1\|,\ldots,\|y_n\| \leq 1\} .$$

This Banach space may be identified with the Banach space

$$L(E_1,\ldots,L(E_{n-1},L(E_n,F))\ldots)$$

under the identification which takes an element $A$ of the latter into the form given by

$$A(y_1,\ldots,y_n) = ((\ldots(Ay_1)\ldots)y_{n-1})y_n \ .$$

If $E_1 = \ldots = E_n = E$, we abbreviate $L(E_1 \times\ldots\times E_n,F)$ by $L^n(E,F)$. The set of symmetric elements of it is denoted by $L^n_{sym}(E,F)$. If $A$ is in $L(E_1 \times\ldots\times E_n,F)$ we denote the value $A(y_1,\ldots,y_n)$ by $Ay_1 \ldots y_n$. Also, if $y$ is in $E$ then $y^n$ means $(y,\ldots,y)$ $n$ times, so that $Ay^n$ is defined if $A$ is in $L^n(E,F)$. If $f\colon U \longrightarrow F$ (with $U$ open in $E$) is $k$ times differentiable then $D^k f$ takes values in $L^n(E,F)$.

<u>Theorem</u> 1 (<u>product rule</u>). <u>Let</u> $E$, $F_1$, $F_2$, <u>and</u> $G$ <u>be Banach spaces, let</u> $U$ <u>be open in</u> $E$, <u>let</u> $f\colon U \longrightarrow F_1$ <u>and</u> $g\colon U \longrightarrow F_2$ <u>be</u> $C^k$, <u>let</u> $(z_1,z_2) \rightsquigarrow z_1\cdot z_2$ <u>be in</u> $L(F_1 \times F_2,G)$ <u>and define</u> $f\cdot g$ <u>by</u> $(f\cdot g)(x) = f(x)\cdot g(x)$. <u>Then</u> $f\cdot g\colon U \longrightarrow G$ is $C^k$ <u>and</u>

$$D(f\cdot g)(x)y = Df(x)y\cdot g(x) + f(x)\cdot Dg(x)y \ . \tag{1}$$

<u>Proof</u>. Suppose $f$ and $g$ are $C^1$. Then

$$f(x+y) = f(x) + Df(x)y + o(y)$$

$$g(x+y) = g(x) + Dg(x)y + o(y)$$

so that

$$f(x+y)\cdot g(x+y) = f(x)\cdot g(x) + Df(x)y\cdot g(x) + f(x)\cdot Dg(x)y + o(y) \ .$$

Thus $f\cdot g$ is $C^1$ and (1) holds, so that the theorem is proved for $k = 1$ Suppose the theorem to be true for $k-1$, and let $f$ and $g$ be $C^{k-1}$.

Then (1) holds. The mapping $\mu: L(E,F_1) \times F_2 \longrightarrow L(E,G)$ given by $(A,z) \rightsquigarrow B$, where $By = Ay \cdot z$, is continuous and bilinear. Now $Df$ and $g$ are $C^{k-1}$, so by the theorem for $k-1$, $x \rightsquigarrow \mu(Df(x),g(x))$ is $C^{k-1}$, and similarly for the other term. Therefore $D(f \cdot g)$ is $C^{k-1}$, so $f \cdot g$ is $C^k$. This concludes the proof.

The same proof shows that the theorem with "$C^k$" replaced by "k times differentiable" is true.

Theorem 2 (chain rule). *Let* $E$, $F$, *and* $G$ *be Banach spaces, let* $U$ *be open in* $E$, *let* $V$ *be open in* $F$, *and let* $f: U \longrightarrow V$ *and* $g: V \longrightarrow G$ *be* $C^k$. *Then* $g \circ f$ *is* $C^k$ *and*

$$D(g \circ f)(x) = Dg(f(x))Df(x) . \tag{2}$$

Proof. Suppose $f$ and $g$ are $C^1$. Then

$$f(x+y) = f(x) + Df(x)y + o(y) ,$$

$$\begin{aligned}(g \circ f)(x+y) &= g(f(x+y)) \\ &= g(f(x)) + Dg(f(x))(Df(x)y + o(y)) + o(Df(x)y + o(y)) \\ &= g(f(x)) + Dg(f(x))Df(x)y + o(y) .\end{aligned}$$

Hence $g \circ f$ is $C^1$ and (2) holds. Thus the theorem holds for $k = 1$. Suppose the theorem to be true for $k-1$, and let $f$ and $g$ be $C^k$. Then $Dg$ and $f$ are $C^{k-1}$, so $Dg \circ f$ is $C^{k-1}$. Also $Df$ is $C^{k-1}$. The mapping of $L(F,G) \times L(E,F)$ into $L(E,G)$ which takes two linear operators into their product is continuous and bilinear, so by Theorem 1, $(Dg \circ f)(Df)$ is $C^{k-1}$. By (2), therefore, $D(g \circ f)$ is $C^{k-1}$ and $g \circ f$ is $C^k$, which completes the proof.

The following formulas are easily proved by induction, for $C^k$

functions $f$ and $g$ :

$$D^k(f\cdot g)(x)y_1 \ldots y_k = \sum_{q=0}^{k} \Sigma\, D^q f(x)y_{i_1} \ldots y_{i_q} \cdot D^{k-q}g(x)y_{j_1} \ldots y_{j_{k-q}} ,$$

where the inner sum is over all $\binom{k}{q}$ partitions of $y_1,\ldots,y_k$ into two sets with $i_1 <\ldots< i_q$ and $j_1 <\ldots< j_{k-q}$ , and

$$D^k(g\circ f)(x)y_1\ldots y_k =$$

$$\sum_{q=1}^{k} \Sigma\, D^q g(f(x))D^{r_1}g(x)y_1^{(1)}\ldots y_{r_1}^{(1)}D^{r_2}g(x)y_1^{(2)}\ldots y_{r_2}^{(2)}\ldots D^{r_q}g(x)y_1^{(q)}\ldots y_{r_q}^{(q)},$$

where the inner sum is over all $k!/r_1!\ldots r_q!$ partitions of $y_1,\ldots,y_k$ into $q$ sets with $r_1,r_2,\ldots,r_q$ elements and the natural ordering in each set.

Let us define

$$(D^q f\cdot D^{k-q}g)(x)y_1\ldots y_k = D^q f(x)y_1\ldots y_q \cdot D^{k-q}g(x)y_{q+1}\ldots y_k$$

and

$$(D^{r_1}g\cdot D^{r_2}g\ldots D^{r_q}g)(x)y_1\ldots y_k =$$

$$D^{r_1}g(x)y_1\ldots y_{r_1} D^{r_2}g(x)y_{r_1+1}\ldots y_{r_1+r_2}\ldots D^{r_q}g(x)y_{r_1+\ldots r_{q-1}+1}\ldots y_k .$$

We shall see later that if $f$ is $C^k$ then $D^k f$ is symmetric. Let us denote by Sym the symmetrizing operator; that is, if $\varphi \in L^k(E,F)$ then Sym $\varphi$ in $L^k_{sym}(E,F)$ is defined by

$$(\mathrm{Sym}\ \varphi)(y_1,\ldots,y_k) = \frac{1}{k!}\sum_{\pi} \varphi(y_{\pi(1)},\ldots,y_{\pi(k)}) ,$$

where the summation is over all permutations $\pi$ of $1,\ldots,k$ . Then we may write

$$D^k(f\cdot g) = \mathrm{Sym} \sum_{q=0}^{k} \binom{k}{q} D^q f \cdot D^{k-q} g \ ,$$

$$D^k(g\circ f) = \mathrm{Sym} \sum_{q=1}^{k} \sum_{r_1+\ldots+r_q=k} \frac{k!}{r_1!\ldots r_q!} (D^q f)\circ g \cdot D^{r_1} g \cdot D^{r_2} g \ldots D^{r_q} g \ .$$

(The formulas on p.3 of [6] should be corrected to take symmetrization into account.)

The following is another proof of a theorem of Abraham [6, p.6]. By $o(y^k)$ we mean a function such that $o(y^k)/\|y\|^k \longrightarrow 0$ as $y \longrightarrow 0$ with $y \neq 0$.

_Theorem 3 (converse of Taylor's theorem). Let E and F be Banach spaces, let U be open in E, and suppose that $f\colon U \longrightarrow F$ satisfies_

$$f(x+y) = a_0(x) + a_1(x)y + \frac{a_2(x)}{2!} y^2 + \ldots + \frac{a_k(x)}{k!} y^k + o(y^k) \tag{3}$$

_where the $a_j(x)$ are in $L^j_{sym}(E,F)$ and each $a_j$ is continuous. Then f is $C^k$ and $a_j = D^j f$ for $j = 0,1,\ldots,k$._

_Proof._ For $k = 1$ this is the definition. Suppose the theorem is true for $k-1$. Then in (3), since $(a_k(x)/k!)y^k = o(y^{k-1})$, we know that $a_j = D^j f$ for $j = 0,1,\ldots,k-1$. Now let us expand $f(x+y+z)$ in two different ways:

$$f(x+y+z) = f(x+y) + Df(x+y)z + \ldots + \frac{1}{(k-1)!} D^{k-1} f(x+y) z^{k-1} + \frac{a_k(x+y)}{k!} z^k + o(z^k) \ ,$$

$$f(x+y+z) = f(x) + Df(x)(y+z) + \ldots + \frac{1}{(k-1)!} D^{k-1} f(x)(y+z)^{k-1} + \frac{a_k(x)}{k!} (y+z)^k + o((y+z)^k) \ .$$

Fix $x$ and restrict $z$ so that $\frac{1}{4}\|y\| \le \|z\| \le \frac{1}{2}\|y\|$ . Then it does not matter whether we write $o(z^k)$, $o((y+z)^k)$, or $o(y^k)$ . Subtract the two equations, collecting coefficients of $z$ and denoting the coefficient of $z^j$ by $g_j(y)$ . Then

$$(4) \qquad g_0(y) + g_1(y)z + \ldots + g_{k-1}(y)z^{k-1} + g_k(y)z^k = o(y^k) .$$

Now

$$g_k(y)z^k = \frac{1}{k!}[a_k(x+y) - a_k(x)]z^k ,$$

and by the continuity of $a_k$ this is $o(y^k)$ , so we may drop this term. We claim that each term separately in (4) is $o(y^k)$ . To see this, let $\lambda_1,\ldots,\lambda_k$ be distinct numbers, and replace $z$ by $\lambda_i z$ for $i=1,\ldots k$ . In this way we obtain $k$ equations which we write as

$$\begin{pmatrix} 1 & \lambda_1 & \cdots & \lambda_1^{k-1} \\ 1 & \lambda_2 & \cdots & \lambda_2^{k-1} \\ \vdots & \vdots & & \vdots \\ 1 & \lambda_k & \cdots & \lambda_k^{k-1} \end{pmatrix} \begin{pmatrix} g_0(y) \\ g_1(y)z \\ \vdots \\ g_{k-1}(y)z^{k-1} \end{pmatrix} = \begin{pmatrix} o(y^k) \\ o(y^k) \\ \vdots \\ o(y^k) \end{pmatrix}$$

Since the $\lambda_i$ are distinct, the matrix is invertible (it is the Vandermonde matrix with determinant $\prod_{i<j} (\lambda_j - \lambda_i))$ . Therefore each $g_j(y)z^i$ is $o(y^k)$ . In particular, this is true for $j = k-1$ . But (and here we use the symmetry of $a_k$)

$$g_{k-1}(y)z^{k-1} = \left[\frac{D^{k-1}f(x+y)}{(k-1)!} - \frac{D^{k-1}f(x)}{(k-1)!} - \frac{ka_k(x)y}{k!}\right]z^{k-1} .$$

Therefore the term in brackets is $o(y)$ . By definition of the derivative, this means that $D^k f(x) = a_k(x)$ , and since $a_k$ is continuous, $f$ is $C^k$ . This completes the proof.

So far what we have done would be valid in the more general context of a normed linear space over a valued field of characteristic $0$. Consider however the function $f\colon \mathbb{Q} \longrightarrow \mathbb{Q}$ defined as follows. Let $\varepsilon_n$ be a sequence of irrational numbers decreasing to $0$, and let $f(0) = 0$,

$$f(x) = a_n ; \quad \varepsilon_n < |x| < \varepsilon_{n+1} , \quad x \in \mathbb{Q} ,$$

where the $a_n$ are rational numbers so chosen that $f(x) = o(x)$ but not $f(x) = o(x^2)$. Then $f$ is $C^\infty$ as a function from $\mathbb{Q}$ to $\mathbb{Q}$ (in the sense of definitions analogous to those given above for $\mathbb{R}$) since $Df = 0$, but Taylor's theorem is not satisfied at $x = 0$. Also, $f$ is not locally constant even though $Df = 0$. (A function $f$ on a topological space is locally constant in case every point $x$ has a neighborhood $V$ such that $f(y) = f(x)$ for all $y$ in $V$. It follows that a locally constant function is constant on each connected component of the space.) It is not the incompleteness of $\mathbb{Q}$ which causes the trouble in the above example, but the fact that $\mathbb{Q}$ is not locally connected. To proceed further we must make substantial use of the fact that we are working over the real number field.

*Theorem 4. Let* $E$ *and* $F$ *be Banach spaces, let* $U$ *be open in* $E$, *let* $f\colon U \longrightarrow F$ *be differentiable, and suppose that* $Df = 0$. *Then* $f$ *is locally constant.*

*Proof.* Let $x$ be in $U$ and let $a > 0$ be so small that the open ball $V$ with center $x$ and radius $a$ is contained in $U$. Let $x+y$ be in $V$ and let $\varphi\colon [0,1] \longrightarrow V$ be defined by $\varphi(t) = x+ty$. Thus $\varphi$ is the line segment joining $x$ and $x+y$. Let $\varepsilon > 0$ and let

$$S_\varepsilon = \{t \in [0,1]: \|f(x) - f(\varphi(s))\| \le \varepsilon s \quad \text{for} \quad 0 \le s \le t\} .$$

This is a closed set containing $0$. It is also open, for if $t_0 \in S_\varepsilon$ then

$$\begin{aligned} f(\varphi(t_0+h)) &= f(x+(t_0+h)y) = f(x+t_0y) + hDf(x+t_0y)y + o(hy) \\ &= f(\varphi(t_0)) + o(h) , \end{aligned}$$

and by the triangle inequality, $\|f(x) - f(\varphi(t_0+h))\| \le \varepsilon(t_0+h)$ for $h$ small enough. Since $[0,1]$ is connected (this is an immediate consequence of the least upper bound property of $\mathbb{R}$) it follows that $S_\varepsilon = [0,1]$. Therefore $\|f(x) - f(x+y)\| \le \varepsilon$. Since $\varepsilon$ is arbitrary, $f(x) = f(x+y)$. QED

The quickest approach to integration of continuous functions is the following (see [4]). Let $I = [a,b]$, $-\infty < a < b < \infty$, and let $F$ be a Banach space. A <u>step function</u> $f: I \longrightarrow F$ is a function which for some partition $a = a_0 < a_1 < \ldots < a_n = b$ is constant on each interval $(a_i, a_{i+1})$. If $f: I \longrightarrow F$ is a step function, define $\int_a^b f(t)dt$ in the obvious way. Let $\|f\| = \sup\{\|f(t)\|: a \le t \le b\}$, and let $\mathcal{R}$ be the completion of the step functions in this norm. An element of $\mathcal{R}$ is a function, since uniform convergence implies pointwise convergence. A function in $\mathcal{R}$ is called a <u>regulated</u> function from $I$ to $F$. A proof quite analogous to the proof of Theorem 4 shows that every continuous function from $I$ to $F$ is regulated. Since $\|\int_a^b f(t)dt\| \le (b-a)\|f\|$, the linear functional $f \leadsto \int_a^b f(t)dt$ extends by continuity to $\mathcal{R}$. Thus we have defined the integral of every regulated, and in particular every continuous, function from $I$ to $F$. This is not quite as general as the Riemann integral, which is defined for some non-regulated functions, but if a more general integral is needed it is preferable to develop the Bochner integral (which is the Lebesgue integral if $F = \mathbb{R}$).

If $I$ is an open subset of $\mathbb{R}$ and $f: I \longrightarrow E$ is $C^1$, then $Df(t)$, for $t$ in $I$, is in $L(\mathbb{R},E)$. Thus $Df(t)1$ is an element of $E$, and it is simply the ordinary derivative

$$\frac{df}{dt}(t) = f'(t) = \lim_{h \to 0} \frac{f(t+h) - f(t)}{h} .$$

If $f: I \longrightarrow E$ is continuous and $a$ is in $I$, we claim that

$$g(t) = \int_a^t f(s)ds$$

is $C^1$ and $g' = f$. To see this, observe that

$$\int_a^{t+h} f(s)ds - \int_a^t f(s)ds$$

$$= \int_t^{t+h} f(s)ds$$

$$= hf(t) + \int_t^{t+h} (f(s) - f(t))ds$$

$$= hf(t) + o(h)$$

by the continuity of $f$.

<u>Theorem</u> 5 (<u>fundamental theorem of calculus</u>). <u>Let</u> $E$ <u>and</u> $F$ <u>be Banach spaces, let</u> $U$ <u>be open in</u> $E$, <u>let</u> $f: U \longrightarrow F$ <u>be</u> $C^1$, <u>and let</u> $x+ty$ <u>be in</u> $U$ <u>for</u> $0 \leq t \leq 1$. <u>Then</u>

$$f(x+y) = f(x) + \int_0^1 Df(x+ty)ydt .$$

<u>Proof.</u> Define $\varphi$, for $0 \leq t \leq 1$, by $\varphi(t) = f(x+ty)$. For $0 < t < 1$, $\varphi'(t) = Df(x+ty)y$ by the chain rule. Define $\psi$, for $0 \leq t \leq 1$, by

$$\psi(t) = f(x) + \int_0^t Df(x+sy)y\,ds .$$

Then for $0 < t < 1$, $\psi'(t) = Df(x+ty)y$ by the continuity of $Df$. By Theorem 4, $\varphi-\psi$ is constant on $(0,1)$, and since $\varphi$ and $\psi$ are continuous on $[0,1]$, $\varphi-\psi$ is constant on $[0,1]$. Since $\varphi(0) = \psi(0)$, $\varphi(1) = \psi(1)$. QED

An immediate corollary is the <u>mean value theorem</u>: under the hypotheses of Theorem 5,

$$\|f(x+y) - f(x)\| \leq \sup_{0 \leq t \leq 1} \|Df(x+ty)\| \|y\| .$$

A function $f\colon U \longrightarrow F$ is called <u>Lipschitz</u> in case for some constant $\kappa < \infty$ (called a <u>Lipschitz constant</u> for $f$)

$$\|f(x_1) - f(x_2)\| \leq \kappa\|x_1 - x_2\| ; \qquad x_1, x_2 \in U .$$

It is called <u>locally Lipschitz</u> in case for each $x$ in $U$ there is a neighborhood $V$ of $x$ in $U$ such that the restriction of $f$ to $V$ is Lipschitz. Thus a $C^1$ function is locally Lipschitz.

<u>Theorem</u> 6 (<u>Taylor's theorem</u>). <u>Let</u> $E$ <u>and</u> $F$ <u>be Banach spaces, let</u> $U$ <u>be open in</u> $E$, <u>and let</u> $f\colon U \longrightarrow F$ <u>be</u> $C^k$. <u>Then</u>

$$f(x+y) = f(x) + Df(x)y + \ldots + \frac{D^k f(x)}{k!} y^k + o(y^k) .$$

<u>Proof.</u> For $k = 1$ this is true. Suppose it is true for $k-1$ and let $f$ be $C^k$. Then $Df$ is $C^{k-1}$, so that

$$Df(x+ty)y = Df(x)y + D^2 f(x)ty^2 + \ldots + \frac{D^k f(x)}{(k-1)!} t^{k-1} y^k + o(t^{k-1}y^k) .$$

Integrate this between 0 and 1 and apply Theorem 5. QED

Theorem 7. Let $E$ and $F$ be Banach spaces, let $U$ be open in $E$, and let $f: U \longrightarrow F$ be $C^k$. Then $D^j f$ is symmetric, for $j = 0,1,\dots,k$.

Proof. Let $a_j(x) = \operatorname{Sym} D^j f(x)$. Since $y^j$ is symmetric, $a_j(x)y^j = D^j f(x)y^j$. By Taylor's theorem,

$$f(x+y) = a_0(x) + a_1(x)y + \dots + \frac{a_k(x)}{k!} y^k + o(y^k) .$$

By the converse of Taylor's theorem (Theorem 3), $a_j(x) = D^j f(x)$. Therefore $D^j f(x)$ is symmetric. QED

## 2. Picard's method

We shall be studying non-linear time-independent differential equations. In doing so, however, it will be useful to have some information about linear time-dependent differential equations.

Recall that the differential equation

$$\frac{df(t)}{dt} = g(t) ,$$

with $g$ a continuous function of $t$, is solved by integration:

$$f(t) = f(t_0) + \int_{t_0}^{t} g(s)ds ,$$

and we have sketched how the integral may be defined. In very close analogy, the linear time-dependent equation may be solved by the product integral. This is an ancient device going back at least to Volterra at the turn of the century, but it keeps being rediscovered.

If $E$ is a Banach space and $A$ is in $L(E)$ we define $e^A$ by

$$e^A = \sum_{n=0}^{\infty} \frac{A^n}{n!} .$$

This series is absolutely convergent in $L(E)$ and we have the crude estimate

$$\|e^A\| \leq e^{\|A\|} .$$

For $t$ and $s$ in $\mathbb{R}$,

$$e^{tA}e^{sA} = e^{(t+s)A} ,$$

so in particular each $e^{tA}$ is invertible with inverse $e^{-tA}$. (However, $e^Ae^B$ is not in general equal to $e^{A+B}$ unless $A$ and $B$ commute.) The operators $e^{tA}$ form a one-parameter group, and

$$\frac{d}{dt} e^{tA} = Ae^{tA} .$$

Thus if $x$ is in $E$, $\xi(t) = e^{tA}x$ is the solution of the linear time-independent differential equation

$$\frac{d\xi(t)}{dt} = A\xi(t)$$

with initial condition $\xi(0) = x$.

We claim that for $A$ and $B$ in $L(E)$,

$$(1) \qquad \|e^A - e^B\| \leq e^{\max\{\|A\|,\|B\|\}}\|A-B\| .$$

To see this, write $A^n - B^n$ as the telescoping sum

$$A^n - B^n = A^{n-1}(A-B) + A^{n-2}(A-B)B + \ldots + A(A-B)B^{n-2} + (A-B)B^{n-1}$$

Thus

$$\|A^n - B^n\| \leq n(\max\{\|A\|,\|B\|\})^{n-1}\|A-B\| ,$$

and so (1) holds.

Let $I = [a,b]$ with $-\infty < a < b < \infty$, let $E$ be a Banach space, and let $\mathcal{R}$ be the Banach space of all regulated functions $A$ from $I$ to $L(E)$, in the norm

$$\|A\| = \sup_{a \leq t \leq b} \|A(t)\| .$$

We wish to define the product integral, which we will denote by

$$\prod_a^b (1 + A(t)dt) ,$$

for $A$ in $\mathcal{R}$.

If $A$ is a step function, $A(t) = A_j$ for $t_{j-1} < t < t_j$ where $a = t_0 < t_1 < \ldots < t_{n-1} < t_n = b$, set $\Delta t_j = t_j - t_{j-1}$, and define

$$\prod_a^b (1 + A(t)dt) = e^{\Delta t_n A_n} \cdots e^{\Delta t_1 A_1} .$$

Notice that the operators with the smallest value of the time parameter operate first. If $B$ is another step function we claim that

$$(2) \qquad \left\|\prod_a^b (1 + A(t)dt) - \prod_a^b (1 + B(t)dt)\right\| \leq$$

$$e^{(b-a)\max\{\|A\|,\|B\|\}}(b-a)\|A-B\| .$$

To prove this, assume that we have a common refinement of the two partitions of $[a,b]$, write the difference of the two product integrals as a telescoping sum, and estimate using (1).

By (2), the mapping

$$A \rightsquigarrow \prod_a^b (1 + A(t)dt)$$

is uniformly continuous on each bounded set in the space of step functions, and so has a unique continuous extension (denoted in the same way) to all of $\mathcal{Q}$ , which by definition is the completion of the space of step functions. In particular, the product integral is defined for all continuous $A: I \longrightarrow L(E)$ . The estimate (2) extends by continuity to all $A$ and $B$ in $\mathcal{Q}$ .

We mention in passing that the product integral is what is frequently called the time-ordered exponential, denoted by

$$\widetilde{\mathcal{T}}\, e^{\int_a^b A(t)dt} .$$

(This notation is somewhat abusive, since $\int_a^b A(t)dt$ and $\int_a^b B(t)dt$ may be equal without the corresponding time-ordered exponentials being equal.) It may be defined, by power series expansion, to be

$$1 + \int_a^b A(t_1)dt_1 + \iint_{a\leq t_1\leq t_2\leq b} A(t_2)A(t_1)dt_1dt_2 + \dots .$$

Notice that operators with the smallest value of the time parameter always operate first. *There are no factorials in this expansion since* the restriction $a \leq t_1 \leq \dots \leq t_n \leq b$ reduces the domain of integration to $1/n!$ of what it would be otherwise. It is not hard to show that

$$\prod_a^b (1 + A(t)dt) = \widetilde{\mathcal{T}}\, e^{\int_a^b A(t)dt} ,$$

but we omit the proof.

It is easy to see that the value of any product integral is an invertible operator. If $a > b$ we define

$$\prod_a^b (1 + A(t)dt) = \{\prod_b^a (1 + A(t)dt\}^{-1} .$$

With this convention we always have

(3) $$\prod_b^c (1+A(t)dt) \prod_c^b (1+A(t)dt) = \prod_a^c (1+A(t)dt) .$$

We claim that if $a < t < b$ and $A$ is continuous, then

(4) $$\frac{d}{dt} \prod_a^t (1+A(s)ds) = A(t) \prod_a^t (1+A(s)ds) .$$

To prove this we need only show that

(5) $$\prod_t^{t+h} (1+A(s)ds) = 1+A(t)h + o(h) ,$$

for then (4) follows by (3) and the definition of derivative. To prove (5), we use the estimate (2) and find

$$\left\| \prod_t^{t+h} (1+A(s)ds) - \prod_t^{t+h} (1+A(t)ds) \right\|$$
$$\leq e^{h\|A\|} h \sup_{t \leq s \leq t+h} \|A(s) - A(t)\| = o(h) ,$$

by the continuity of $A$ and the fact that $t$ and $A(t)$ are fixed. But

$$\prod_t^{t+h} (1+A(t)ds) = e^{hA(t)} = 1+A(t)h + o(h) .$$

Thus (4) is true.

Theorem 1. *Let* $I$ *be an open interval, let* $E$ *be a Banach space, and let* $A: I \to L(E)$ *be continuous. For all* $x$ *in* $E$ *and* $t_0$ *in* $I$ *there is a unique* $C^1$ *function* $\xi: I \to E$ *with* $\xi(t_0) = x$ *and satisfying*

(6) $$\frac{d\xi(t)}{dt} = A(t)\xi(t) .$$

*It is given by*

(7) $$\xi(t) = \prod_{t_0}^t (1+A(s)ds)x .$$

*If* $B: I \longrightarrow L(E)$ *is also continuous and we let* $\eta: I \longrightarrow E$ *be the solution of*

$$\frac{d\eta(t)}{dt} = B(t)\eta(t)$$

*with* $\eta(t_0) = x$ *then for all* $t$ *in* $I$,

$$(8) \qquad \|\xi(t) - \eta(t)\| \le e^{|t-t_0|\max\{\|A\|,\|B\|\}}|t-t_0|\|A-B\|\|x\| ,$$

*where* $\|A\|$ *denotes the supremum of* $\|A(s)\|$ *for* $s$ *between* $t_0$ *and* $t$.

*Proof.* We have just seen that (7) solves (6). The uniqueness is proved in the usual way: if $\tilde{\xi}$ is also a solution with $\tilde{\xi}(t_0) = x$ then for each $\varepsilon > 0$ the set of $t$ such that $\|\xi(s) - \tilde{\xi}(s)\| \le \varepsilon|s-t_0|$ for all $s$ between $t_0$ and $t$ is both open and closed, and contains $t_0$, and so is all of $I$. Since $\varepsilon$ is arbitrary, $\tilde{\xi} = \xi$. The inequality (8) follows immediately from (2). QED

Picard's method for proving the local existence and uniqueness of solutions to systems of ordinary differential equations is based on the following simple fixed point theorem.

*Theorem* 2 (*fixed point theorem for proper contractions*). *Let* $M$ *be a complete non-empty metric space, and let* $\Phi: M \longrightarrow M$ *be such that for some* $a < 1$,

$$d(\Phi(x_1),\Phi(x_2)) \le ad(x_1,x_2) ; \qquad x_1,x_2 \in M ,$$

*where* $d$ *is the metric on* $M$. *Then there exists a unique fixed point* $x_0$ *for* $\Phi$. *If* $x_1$ *is any element of* $M$ *then* $\Phi^n(x_1) \longrightarrow x_0$.

*Proof.* Let $x_1$ be in $M$. Then

$$d(\Phi^{n+1}(x_1),\Phi^n(x_1)) \leq a^n d(\Phi(x_1),x_1) .$$

Therefore, by the triangle inequality,

$$d(\Phi^{n+k}(x_1),\Phi^n(x_1)) \leq d(\Phi^{n+k}(x_1),\Phi^{n+k-1}(x_1)) + \dots$$
$$+ d(\Phi^{n+1}(x_1),\Phi^n(x_1)) \leq \left(\sum_{j=n}^{n+k-1} a^j\right) d(\Phi(x_1),x_1) .$$

Therefore $\Phi^n(x_1)$ is a Cauchy sequence. Let $x_0$ be its limit. Since $\Phi$ is continuous,

$$\Phi(x_0) = \Phi \lim_n \Phi^n(x_1) = \lim_n \Phi^{n+1}(x_1) = \lim_n \Phi^n(x_1) = x_0 ,$$

and $x_0$ is a fixed point. The uniqueness is obvious. QED

Before applying this theorem to differential equations, let us use it to prove the inverse function theorem (following Lang [5, p.12]). A $C^k$ <u>diffeomorphism</u> of an open set $U$ in a Banach space $E$ to an open set $V$ in a Banach space $F$ is a bijective $C^k$ map $f: U \longrightarrow V$ such that $f^{-1}$ is $C^k$. If $U$ is non-empty and $f: U \longrightarrow V$ is a $C^k$ diffeomorphism $(k \geq 1)$ then $E$ and $F$ are isomorphic Banach spaces (not necessarily isometric), for if $x$ is in $U$ then by the chain rule $Df(x)$ and $Df^{-1}(f(x))$ are inverse continuous linear transformations between $E$ and $F$. A <u>local</u> $C^k$ <u>diffeomorphism at</u> $x$ in $U$ is a map $f$ defined in a neighborhood of $x$ such that for some open neighborhood $W$ of $x$, the restriction of $f$ to $W$ is a $C^k$ diffeomorphism of $W$ to $f(W)$.

<u>Theorem 3 (inverse function theorem)</u>. <u>Let</u> $E$ <u>and</u> $F$ <u>be Banach spaces,</u> $U$ <u>open in</u> $E$, $x_0$ <u>in</u> $U$, $f: U \longrightarrow F$ <u>of class</u> $C^k$,

$k \geq 1$ . <u>Suppose that</u> $Df(x_0)$ <u>is invertible. Then</u> $f$ <u>is a local</u> $C^k$ <u>diffeomorphism at</u> $x$ .

<u>Proof.</u> We may replace $f$ by $Df(x_0)^{-1} \circ f$ , and so we may assume without loss of generality that $E = F$ and $Df(x_0) = 1$ . Also, we may assume without loss of generality that $x_0 = f(x_0) = 0$ .

Now let

$$g(x) = x - f(x) .$$

Then $Dg(0) = 1 - 1 = 0$ . By continuity there is an $r > 0$ such that

$$\|Dg(x)\| \leq \frac{1}{2} , \qquad \|x\| \leq 2r .$$

By the mean value theorem, $\|g(x)\| \leq \frac{1}{2}\|x\|$ for $\|x\| \leq 2r$ . Let $B_s$ be the closed ball of center 0 and radius $s$ . Then

$$g\colon B_{2r} \longrightarrow B_r .$$

Let $y$ be in $B_r$ . We claim that there is a unique $x$ in $B_{2r}$ such that $f(x) = y$ . To see this, let

$$h(x) = y + x - f(x) = y + g(x) .$$

Then $h\colon B_{2r} \longrightarrow B_{2r}$ , and by the mean value theorem

$$\|h(x_1) - h(x_2)\| = \|g(x_1) - g(x_2)\| \leq \frac{1}{2}\|x_1 - x_2\|$$

for $x_1, x_2$ in $B_{2r}$ . By the fixed point theorem for proper contractions, $h$ has a unique fixed point in $B_{2r}$ ; that is, there is a unique $x$ in $B_{2r}$ such that $f(x) = y$ . Therefore

$$\varphi = f^{-1}\colon B_r \longrightarrow B_{2r}$$

is well-defined. Since $x = g(x) - f(x)$ ,

$$\|x_1 - x_2\| \leq \|g(x_1) - g(x_2)\| + \|f(x_1) - f(x_2)\|$$
$$\leq \frac{1}{2}\|x_1 - x_2\| + \|f(x_1) - f(x_2)\| ,$$

so that

$$\|x_1 - x_2\| \leq 2\|f(x_1) - f(x_2)\| ; \qquad x_1, x_2 \in B_{2r} .$$

Thus $\varphi\colon B_r \longrightarrow B_{2r}$ is Lipschitz.

We claim that on a Banach space $E$, the invertible elements of $L(E)$ are an open set, and on this open set the function $A \rightsquigarrow A^{-1}$ is $C^\infty$. To see this, let $A$ be invertible and suppose that $\|B\| < 1/\|A^{-1}\|$ Then $A+B = A(1 + A^{-1}B)$, so that

$$(A+B)^{-1} = \left( \sum_{n=0}^{\infty} (-1)^n (A^{-1}B)^n \right) A^{-1} .$$

Since the power series is convergent, the function is certainly $C^\infty$.

Consequently, if $x$ is small enough, $Df(x)$ is invertible, since $Df(0) = 1$. We assume that we have chosen $r$ small enough so that $Df(x)$ is invertible for $\|x\| \leq 2r$ with $\|Df(x)^{-1}\| \leq c$ for some $c$.

We claim that $\varphi$ is differentiable on the interior of $B_r$. To see this, let $\|y\| < r$, $\|x\| \leq 2r$, $f(x) = y$, $\|y+y_1\| < r$, $\|x+x_1\| \leq 2r$, $f(x+x_1) = y+y_1$. Then

$$\|\varphi(y+y_1) - \varphi(y) - Df(x)^{-1}y_1\|$$
$$= \|x+x_1-x - Df(x)^{-1}(f(x+x_1) - f(x))\|$$
$$= \|Df(x)^{-1}\{Df(x)x_1 + f(x+x_1) - f(x)\}\|$$
$$\leq c\|f(x+x_1) - f(x) - Df(x)x_1\| = o(x_1)$$

since $f$ is differentiable at $x$. Now $x_1 = \varphi(y+y_1) - \varphi(y)$, and since $\varphi$ is Lipschitz $o(x_1)$ is also $o(y_1)$. Thus $\varphi$ is differentiable and

$$D\varphi = (Df \circ \varphi)^{-1} .$$

Since $A \rightsquigarrow A^{-1}$ is $C^\infty$, $D\varphi$ is a composition of $C^{k-1}$ functions (by induction on $k$) and so is a $C^{k-1}$ function. Thus $\varphi$ is $C^k$ if $f$ is. QED

If $U$ is an open subset of $E$, $X: U \longrightarrow E$ is continuous, and $x$ is in $U$, an <u>integral curve of</u> $X$ <u>starting at</u> $x$ is a $C^1$ function $\xi: I \longrightarrow U$, where $I$ is an open interval containing $0$, such that $\xi(0) = x$ and

$$\frac{d\xi}{dt}(t) = X(\xi(t)) , \qquad t \in I .$$

Thus $\xi$ is an integral curve of $X$ starting at $x$ if and only if $\xi: I \longrightarrow U$ is continuous and

$$\xi(t) = x + \int_0^t X(\xi(s))ds , \qquad t \in I .$$

<u>Theorem</u> 4. <u>Let</u> $U$ <u>be an open subset of a Banach space</u> $E$ <u>and let</u> $X: U \longrightarrow E$ <u>be locally Lipschitz. For each</u> $x$ <u>in</u> $U$ <u>there is an integral curve of</u> $X$ <u>starting at</u> $x$, <u>and any two of them agree on the intersection of their domains of definition. For all</u> $x_0$ <u>in</u> $U$ <u>there is an open neighborhood</u> $V$ <u>of</u> $x_0$, <u>an</u> $a > 0$, <u>and a unique mapping</u>

$$\varphi: (-a,a) \times V \longrightarrow U$$

<u>such that for all</u> $x$ <u>in</u> $V$, $t \rightsquigarrow \varphi(t,x)$ <u>is an integral curve of</u> $X$ <u>starting at</u> $x$. <u>The mapping</u> $\varphi$ <u>is locally Lipschitz. If</u> $X$ <u>is of class</u> $C^k$ <u>so is</u> $\varphi$.

Proof. Let $\xi$ and $\eta$ be two integral curves of $X$ starting at $x$, defined on the interval $I$. Let $J$ be the set of points $t$ in $I$ such that $\xi(t) = \eta(t)$. Then $J$ is clearly closed in $I$ and contains $0$, so we need only show that it is open. Let $t_0$ be in $J$ let $W$ be an open neighborhood of $\xi(t_0) = \eta(t_0)$ on which $X$ is Lipschitz with Lipschitz constant $\kappa$, and choose $\varepsilon > 0$ so small that $\varepsilon\kappa < 1$, that $\xi(t)$ and $\eta(t)$ are in $W$ for $|t-t_0| \leq \varepsilon$, and that $[t_0-\varepsilon, t_0+\varepsilon] \subset I$. For $|t-t_0| \leq \varepsilon$ we have

$$\xi(t) = \xi(t_0) + \int_{t_0}^{t} X(\xi(s))ds ,$$

$$\eta(t) = \eta(t_0) + \int_{t_0}^{t} X(\eta(s))ds ,$$

so that

$$\|\xi(t) - \eta(t)\| \leq \left\|\int_{t_0}^{t} [X(\xi(s)) - X(\eta(s))]ds\right\| \leq \kappa\varepsilon \sup_{|s-t_0|\leq\varepsilon} \|\xi(s) - \eta(s)\| .$$

Since this is true for all $t$ with $|t-t_0| \leq \varepsilon$,

$$\sup_{|t-t_0|\leq\varepsilon} \|\xi(t) - \eta(t)\| \leq \kappa\varepsilon \sup_{|t-t_0|\leq\varepsilon} \|\xi(t) - \eta(t)\| ,$$

and since $\kappa\varepsilon < 1$ this supremum must be $0$, so that $\xi(t) = \eta(t)$ for $|t-t_0| \leq \varepsilon$. This proves the uniqueness of integral curves starting at any point $x$.

Given $x_0$ in $U$, choose a neighborhood $U_0$ of $x_0$ such that $X$ is Lipschitz with some Lipschitz constant $\kappa$ on $U_0$ and bounded with some bound $b$ on $U_0$. Choose an open neighborhood $V$ of $x_0$ in $U_0$ and an $a > 0$ such that $\kappa a < 1$ and

$$ba < \inf_{\substack{x\in V \\ y\notin U_0}} \|x-y\| .$$

Let

$$F = \{y \in E: \inf_{x \in U} \|x-y\| \leq ba\} .$$

Then $F$ is a closed subset of $U_0$ . Let $x$ be in $V$ and let $M$ be the metric space of all continuous mappings $\xi: [-a,a] \longrightarrow F$ such that $\xi(0) = x$ , in the metric

$$d(\xi,\eta) = \sup_{|t| \leq a} \|\xi(t) - \eta(t)\| .$$

This is a complete non-empty metric space. For $\xi$ in $M$ , define $\Phi(\xi)$ by

$$\Phi(\xi)(t) = x + \int_0^t X(\xi(s))ds .$$

Then $\Phi(\xi): [-a,a] \longrightarrow E$ is continuous, $\Phi(\xi)(0) = x$ , and the range of $\Phi(\xi)$ is contained in $F$ , so that $\Phi: M \longrightarrow M$ . The argument used above in proving the uniqueness of integral curves shows that

$$d(\Phi(\xi),\Phi(\eta)) \leq \kappa a d(\xi,\eta) .$$

By Theorem 2, $\Phi$ has a unique fixed point, which is an integral curve for $X$ starting at $x$ .

Let $\varphi(t,x) = \xi(t)$ where $\xi$ is this fixed point. As before,

$$\sup_{|t| \leq a} \|\varphi(t,x_1) - \varphi(t,x_2)\| \leq$$

$$\|x_1-x_2\| + \sup_{|t| \leq a} \|\int_0^t [X(\varphi(s,x_1)) - X(\varphi(s,x_2))]ds\| \leq$$

$$\|x_1-x_2\| + \kappa a \sup_{|t| \leq a} \|\varphi(t,x_1) - \varphi(t,x_2)\| ,$$

so that

$$\sup_{|t| \leq a} \|\varphi(t,x_1) - \varphi(t,x_2)\| \leq \frac{1}{1-\kappa a}\|x_1-x_2\| .$$

Thus $\varphi$ is Lipschitz in $x$, uniformly in $t$ Clearly,

$$\|\varphi(t_1,x) - \varphi(t_2,x)\| = \|\int_{t_1}^{t_2} X(\varphi(s,x))ds\| \leq b|t_1-t_2| ,$$

so that $\varphi$ is also Lipschitz in $t$, uniformly in $x$. Consequently $\varphi$ is Lipschitz in $t$ and $x$ jointly.

It remains to prove that if $X$ is $C^k$ so is $\varphi$. The hard case is $k = 1$; the general case follows by induction.

First, a matter of notation. Suppose that $E_1$, $E_2$, and $F$ are Banach spaces, $U$ is open in $E_1 \oplus E_2$, and $f: U \longrightarrow F$. For each $(x_1,x_2)$ in $U$ we define $D_1f(x_1,x_2)$ to be the derivative (if it exists) of the function $g$ given by $g(x) = f(x,x_2)$, evaluated at the point $x_1$. The function $D_1f$ is called a partial derivative of $f$, and $D_2f$ is defined similarly. It is easy to see that $f$ is $C^1$ if and only if $D_1f$ and $D_2f$ exist and are continuous. We apply this now to the case $E_1 = \mathbb{R}$, $E_2 = E$.

Let $X$ be $C^1$. Formally, we expect $D_2\varphi(t,x)$ to satisfy the equation

$$\frac{d}{dt} D_2\varphi(t,x) = DX(\varphi(t,x))\cdot D_2\varphi(t,x) .$$

Let $\psi(t,x)$ be the solution of

(9) $$\frac{d}{dt} \psi(t,x) = DX(\varphi(t,x))\cdot\psi(t,x) , \qquad \psi(0,x) = 1 .$$

This exists and is unique by Theorem 1.

We claim that $\psi$ is continuous. For each $x$, $\psi(t,x)$ is continuous in $t$. Since $X$ is $C^1$, we may choose $U_0$, $a$, and $V$

smaller if necessary so that $DX(\varphi(t,x))$ is bounded uniformly for $|t| \leq a$ and $x$ in $V$. Thus $\psi(t,x)$ is continuous in $t$ uniformly in $x$. Let $x_1$ be fixed in $V$. The set of $\varphi(t,x_1)$ for $|t| \leq a$ is compact. Since $DX$ is continuous, for all $\varepsilon > 0$ there is a $\delta > 0$ such that if

(10) $$\sup_{|t|\leq a} \|\varphi(t,x_1) - \varphi(t,x_2)\| \leq \delta$$

then

(11) $$\sup_{|t|\leq a} \|DX(\varphi(t,x_1)) - DX(\varphi(t,x_2))\| \leq \varepsilon .$$

(To see this, apply the Heine-Borel theorem to the compact set of $\varphi(t,x_1)$ with $|t| \leq a$.) But $\varphi$ is Lipschitz in $s$ uniformly in $t$, so there is a $\delta' > 0$ such that if $\|x_1-x_2\| \leq \delta'$ then (10), and consequently (11), holds. By (8) of Theorem 1, $\psi$ is continuous in $x$. Together with the equicontinuity of $\psi$ in $t$, this shows that $\psi$ is continuous.

Now we want to show that $D_2\varphi(x,t)$ exists and is equal to $\psi(t,x)$. By the fundamental theorem of calculus,

$$\frac{d}{dt}\{\varphi(t,x+y) - \varphi(t,x)\} = X(\varphi(t,x+y)) - X(\varphi(t,x))$$

$$= \int_0^1 DX[\varphi(t,x) + r\{\varphi(t,x+y) - \varphi(t,x)\}]dr\{\varphi(t,x+y) - \varphi(t,x)\} .$$

Let

$$A(t) = DX(\varphi(t,x)) ,$$

$$B_y(t) = \int_0^1 DX[\varphi(t,x) + r\{\varphi(t,x+y) - \varphi(t,x)\}]dr .$$

Then $A$ and $B_y$ are continuous mappings of $[-a,a]$ into $L(E)$, and

$\|A-B_y\| \longrightarrow 0$ as $y \longrightarrow 0$, with $\|B_y\|$ remaining bounded, say $\|A\| \leq c$, $\|B_y\| \leq c$. By (8) of Theorem 1,

$$\sup_{|t|\leq a} \|\psi(t,x)y - \{\varphi(t,x+y) - \varphi(t,x)\}\| \leq e^{ac}a\|A-B_y\|\|y\| = o(y) .$$

Therefore $D_2\varphi(t,x) = \psi(t,x)$. Clearly $\frac{d}{dt}\varphi(t,x) = X(\varphi(t,x))$. Since both $D_2\varphi$ and $D_1\varphi$ exist and are continuous, $\varphi$ is $C^1$.

Now let $X$ be $C^k$, $k \geq 2$. Then $\varphi$ is $C^1$, and as we have seen

$$\frac{d}{dt}\varphi(t,x) = X(\varphi(t,x)) ,$$

$$\frac{d}{dt}\frac{d}{dt}\varphi(t,x) = DX(\varphi(t,x))\cdot X(\varphi(t,x)) ,$$

$$\frac{d}{dt}D_2\varphi(t,x) = DX(\varphi(t,x))\cdot D_2\varphi(t,x) .$$

Since $X$ is $C^k$, the right hand side of this system is a $C^{k-1}$ function of the triple $\varphi$, $\frac{d}{dt}\varphi$, $D_2\varphi$. By induction, this triple is $C^{k-1}$, so $\frac{d}{dt}\varphi$ and $D_2\varphi$ are $C^{k-1}$ and $\varphi$ is $C^k$. QED

Notice that we have shown that if $X$ is $C^k$ then $\varphi$ is $C^{k+1}$ in $t$. A function is said to be $Lip^k$ in case it is $C^{k-1}$ and the $k-1$ derivative is locally Lipschitz. A very easy induction shows that if $X$ is $Lip^k$ so is $\varphi$, thus giving a very simple proof that if $X$ is $C^\infty$ (which is the same as $Lip^k$ for all $k$) so is $\varphi$. It is surprising how difficult it is to show that if $X$ is $C^1$ then $\varphi$ is $C^1$, but there is no simple proof in the literature. For an interesting modern proof, see [12].

## 3. The local structure of vector fields

We have been discussing mappings $X: U \longrightarrow E$ . Loosely speaking, these may be termed vector fields on $U$ . If $R$ is a diffeomorphism of $U$ onto an open set $V$ in the Banach space $F$ then we define the transformed vector field $Y = R_*X$ on $V$ by

$$Y(y) = DR(R^{-1}y)\cdot X(R^{-1}y) .$$

The reason for this definition is as follows. If $X$ is locally Lipschitz it generates a local flow $\varphi(t,x)$ on $U$ with

$$\frac{d}{dt}\varphi(t,x)\Big|_{t=0} = X(x) .$$

This flow may be transported to $V$ by setting

$$(R_*\varphi)(t,y) = R(\varphi(t,R^{-1}y)) .$$

The vector field generating this local flow is the above defined $Y = R_*X$ by the chain rule. If $X$ is $C^k$ and $R$ is $C^k$ then $R_*Y$ is $C^{k-1}$ , and this is the best that can be said in general. However, notice that if $X$ generates a $C^k$ local flow (for example, if $X$ is $C^k$ ) and if $R$ is $C^k$ then the transformed local flow is also $C^k$ , so that $R_*X$ also generates a $C^k$ local flow although it may not be a $C^k$ vector field.

Given a vector field $X$ on $U$ we may seek a diffeomorphism $R$ such that $R_*X$ has a simpler appearance. There is no loss of generality, given $x_0$ in $U$ , in assuming that $E = F$ and $R(x_0) = x_0$ . Thus we are seeking a change of coordinates which simplifies $X$ .

A point $x$ in $U$ is called a <u>regular point</u> of $X$ in case $X(x) \neq 0$, otherwise it is called a <u>critical point</u>. Notice that $x$ is a regular point of $X$ if and only if $R(x)$ is a regular point of $R_*X$. The straightening out theorem asserts that we may choose coordinates in the neighborhood of a regular point so that $X$ becomes a constant. Sternberg's linearization theorem (in the finite dimensional case) asserts that in the neighborhood of a critical point we may, in general but not always, choose coordinates so that $X$ becomes linear. We shall assume throughout this section that $E$ is a finite dimensional Banach space ($\mathbb{R}^s$ with some norm), although the straightening out theorem is true without this assumption.

We shall achieve a considerable simplification in the proof of the Sternberg linearization theorem by using some ideas which are familiar in quantum mechanics. Let $U_0(t)$ and $U(t)$ be two one-parameter groups of transformations on some space (for example, unitary groups on Hilbert space or flows on a manifold) Suppose that the limit

(1) $$\lim_{t \to \infty} U(t)U_0(-t) = W$$

exists. In quantum mechanics this is called the Møller wave operator. There is an analogous operator $W_- = \lim_{t \to -\infty} U(t)U_0(-t)$, and the scattering operator $S$, or S-matrix, of Wheeler and Heisenberg is constructed in terms of these wave operators. Notice that if (1) exists then

$$U(s)WU_0(-s) = W .$$

If $W$ is invertible this means that

$$W^{-1}U(s)W = U_0(s) ,$$

and the two one-parameter groups are conjugate. If they are generated by $X$ and $X_0$ this means that $(W^{-1})_*X = X_0$. In order for $W$ to exist, we see that the flow $U_0(-t)$ must carry points into a region where the flow $U(t)$ is approximately equal to $U_0(t)$; that is, where $X$ is approximately equal to $X_0$.

As a first illustration of this method, we use it to prove the straightening out theorem, although for this simple result the usual proof [6, p.58] is equally easy.

Theorem 1 (straightening out theorem). Let $U$ be an open subset of $\mathbb{R}^s$, let $X: U \longrightarrow \mathbb{R}^s$ be $C^k$ with $k \geq 1$, and let $x_0$ in $U$ be a regular point of $X$. Then there exists a local $C^k$ diffeomorphism $W$ at $x_0$ such that $W_*X$ is constant in a neighborhood of $x_0$.

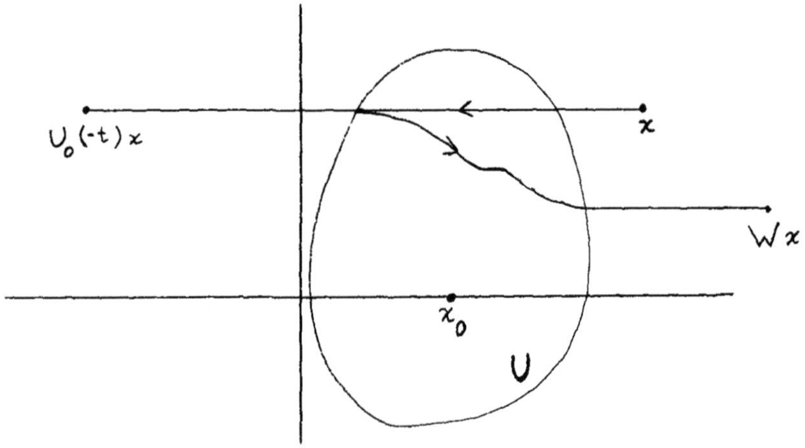


Figure 1. Construction of W (Theorem 1).

Proof. We may assume without loss of generality that $x_0 = X(x_0) = (1,0,\dots,0)$. Let $X_0(x) = (1,0,\dots,0)$ for all $x$ in $\mathbb{R}^s$. Let $U$ be a neighborhood of $x_0$ contained in the half-space $x^1 > 0$ (where $x^1$ denotes the first component of $x$), such that $(X(x))^1 \geq \frac{1}{2}$ on $U$. Let $f$ be a $C^k$ function, $0 \leq f \leq 1$, with support in $U$ which is 1 in a neighborhood of $x_0$. Then

$$\tilde{X} = fX + (1-f)X_0$$

is $C^k$, agrees with $X$ in a neighborhood of $x_0$, agrees with $X_0$ outside $U$, and $(\tilde{X}(x))^1 \geq \frac{1}{2}$ on $U$ by the triangle inequality.

It is clear that $\tilde{X}$ and $X_0$ generate global flows, which we denote by $\tilde{U}(t)$ and $U_0(t)$. Thus $\tilde{U}(t)$ and $U_0(t)$ are $C^k$ diffeomorphisms of $\mathbb{R}^s$ onto itself, for each $t$, and they are one-parameter groups (by the uniqueness assertion of Theorem 4, §2). Let

$$Wx = \lim_{t \to \infty} \tilde{U}(t)U_0(-t)x .$$

It is clear that this limit exists, since $\tilde{U}(t)U_0(-t)x$ is constant in $t$ as soon as $t \geq x^1$. Furthermore,

$$W^{-1}x = \lim_{t \to \infty} U_0(t)\tilde{U}(-t)x$$

exists since $U_0(t)\tilde{U}(-t)x$ is constant in $t$ as soon as $t \geq 2x^1$. Thus $W$ is a $C^k$ diffeomorphism of $\mathbb{R}^s$ onto itself, and clearly

$$W^{-1}\tilde{U}(t)W = U_0(t) .$$

Therefore $(W^{-1})_*\tilde{X} = X_0$. Since $\tilde{X} = X$ in a neighborhood of $x_0$, the proof is complete.

Our problem now is to show that in general we may choose coordinates at a critical point of a vector field so that it becomes linear. The meaning of "in general" will be made clear later. Now we show that this is not always possible; cf. [9, p.812].

Let

$$X\binom{x}{y} = \binom{ax+y^2}{by} .$$

Suppose there is a local diffeomorphism $R$ at $0$ which is $C^2$ and such that $R_*X$ is linear. There is no loss of generality in assuming that $R(0) = 0$ and $DR(0) = 1$. We may write

$$R\binom{x}{y} = \begin{pmatrix} x + \alpha x^2 + \beta xy + \gamma y^2 \\ y + Ax^2 + Bxy + Cy^2 \end{pmatrix} + o\binom{x}{y}^2 ,$$

so that

$$R^{-1}\binom{x}{y} = \begin{pmatrix} x - \alpha x^2 - \beta xy - \gamma y^2 \\ y - Ax^2 - Bxy - Cy^2 \end{pmatrix} + o\binom{x}{y}^2$$

and

$$DR\binom{x}{y} = \begin{pmatrix} 1 + 2\alpha x + \beta y & \beta x + 2\gamma y \\ 2Ax + By & 1 + Bx + 2Cy \end{pmatrix} + o\binom{x}{y} .$$

Notice that this is also equal to $DR \circ R^{-1}\binom{x}{y} + o\binom{x}{y}$. Therefore

$$R_*X\binom{x}{y} = DR \circ R^{-1} \cdot X \circ R^{-1}\binom{x}{y} =$$

$$\begin{pmatrix} 1 + 2\alpha x + \beta y & \beta x + 2\gamma y \\ 2Ax + By & 1 + Bx + 2Cy \end{pmatrix} \begin{pmatrix} a(x - \alpha x^2 - \beta xy - \gamma y^2) + y^2 \\ b(y - Ax^2 - Bxy - Cy^2 \end{pmatrix} + o\binom{x}{y}^2$$

$$= \begin{pmatrix} ax + a\alpha x^2 + \beta bxy + (1 - a\gamma + 2\gamma b)y^2 \\ by + (2Aa - bA)x^2 + Baxy + Cby^2) \end{pmatrix} + o\binom{x}{y}^2 .$$

For this to be linear we must in particular have $(1-a\gamma+2\gamma b) = 0$, and this may be achieved if and only if $a \neq 2b$. Thus the vector field

$$X\binom{x}{y} = \binom{2bx + y^2}{by}$$

cannot be linearized by a $C^2$ change of coordinates.

More generally, consider the orbits of the flow generated by the linear vector field

$$X_0\binom{x}{y} = \binom{kx}{y}$$

for $k > 0$. For $k = 2$, these are sketched in Figure 2. These orbits are branches of the curves $x = cy^k$, $c$ a parameter. If $k$ is an integer, each orbit has a $C^k$ continuation through the origin. This is a property which must be preserved by $C^k$ diffeomorphisms but which can be destroyed by adding a perturbing vector field $X_1$ to $X_0$ with $X_1(0) = 0$ and $DX_1(0) = 0$. Thus for $k$ an integer the local phase portrait is unstable with respect to small perturbations.

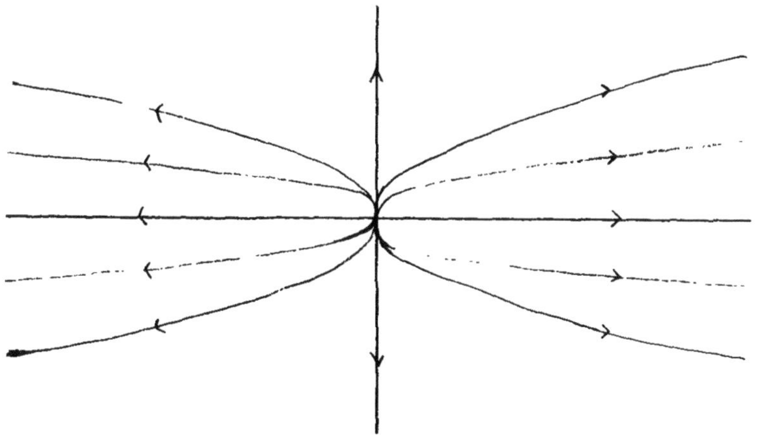

Figure 2. Orbits for $X_0\binom{x}{y} = \binom{2x}{y}$

Thus we see that there may be arithmetic obstructions to linearizing a vector field. We study this problem next.

Let $\mathcal{G}^k$, for $k = 1,2,\ldots,\infty$, be the set of all germs of $C^k$ mappings $R$ at the origin in $\mathbb{R}^s$ leaving the origin fixed and such that $DR(0)$ is invertible (that is, having non-vanishing Jacobian determinant at $0$). This is a group under composition, by the inverse function theorem.

Let $\mathcal{F}^k$, for $k = 1,2,\ldots$, be the set of polynomial mappings $R$ of degree $\leq k$ of $\mathbb{R}^s$ into itself with $R(0) = 0$ and $DR(0)$ invertible. This is a group under composition followed by truncation (throwing away terms of degree $> k$). Let $\mathcal{F}^\infty$ be the set of all formal power series of the form

$$\sum_{j=1}^{\infty} A_j x^j$$

where each $A_j$ is in $L^j_{\mathrm{sym}}(\mathbb{R}^s, \mathbb{R}^s)$ and $A_1$ is invertible This is a group under formal composition.

We have the following commutative diagram of homomorphisms:

$$\begin{array}{ccccccccccc}
\mathcal{G}^\infty & \longrightarrow & \cdots & \longrightarrow & \mathcal{G}^k & \longrightarrow & \mathcal{G}^{k-1} & \longrightarrow & \cdots & \longrightarrow & \mathcal{G}^1 \\
\downarrow & & & & \downarrow & & \downarrow & & & & \downarrow \\
\mathcal{F}^\infty & \longrightarrow & \cdots & \longrightarrow & \mathcal{F}^k & \longrightarrow & \mathcal{F}^{k-1} & \longrightarrow & \cdots & \longrightarrow & \mathcal{F}^1 ,
\end{array}$$

where the mappings in the top row are inclusion, in the bottom row are truncation, and the vertical arrows denote taking Taylor series. (The latter depends on the choice of coordinates.) It is evident, except for $k = \infty$, that $\mathcal{G}^k \to \mathcal{F}^k$ is surjective. The following theorem shows that $\mathcal{G}^\infty \to \mathcal{F}^\infty$ is surjective, too.

Theorem 2. *Let* $A_j$ *be in* $L^j_{sym}(\mathbb{R}^s, \mathbb{R}^s)$ *for* $j = 0,1,2,\ldots$ . *Then there is a* $C^\infty$ *mapping* $F\colon \mathbb{R}^s \to \mathbb{R}^s$ *whose Taylor series at* 0 *is*

$$\sum_{j=0}^{\infty} A_j x^j .$$

Proof. We need only define $F$ in a neighborhood of $0$, for we may extend it to all of $\mathbb{R}^s$ by multiplying with a scalar function $f$ which is 1 in a neighborhood of 0 and has support in the domain of definition of $F$. Let $\alpha\colon \mathbb{R} \to \mathbb{R}$ be $C^\infty$, $0 \le \alpha \le 1$, $\alpha = 1$ in a neighborhood of 0, with support in $[-1,1]$. Let

$$F(x) = \sum_{j=0}^{\infty} A_j x^j \alpha(\|A_j\| \|x\|^2) ,$$

where $\|x\|$ is the Euclidean norm (so that $x \rightsquigarrow \|x\|^2$ is $C^\infty$ ). The j'th term is 0 except where $\|A_j\| \|x\|^2 \le 1$, so that

$$\sum_{j=0}^{\infty} \|A_j\| \|x\|^j \alpha(\|A_j\| \|x\|^2) \le \|A_0\| + \|A_1\| \|x\| + \sum_{j=2}^{\infty} \|x\|^{j-2}$$

and the series converges absolutely for $\|x\| < 1$. In the same way we see that the term-by-term $k$ times differentiated series converges absolutely for $\|x\| < 1$. Therefore $F$ is $C^\infty$ on $\|x\| < 1$. Since $\alpha$ is 1 in a neighborhood of 0, the Taylor series of $F$ at 0 is as stated. QED

Given $T$ in $\mathcal{G}^k$ we may ask whether there is an $R$ in $\mathcal{G}^k$ such that $RTR^{-1}$ is linear. If this is true in $\mathcal{G}^k$ it is also true in $\bar{\mathcal{F}}^k$ since $\mathcal{G}^k \to \bar{\mathcal{F}}^k$ is surjective. The next theorem gives sufficient conditions for linearizing in $\bar{\mathcal{F}}^k$. We remark that "positive" means $\geq 0$. By "eigenvalues" we mean complex eigenvalues.

*Theorem* 3. *Let* $T$ *be in* $\bar{\mathcal{F}}^k$ *for some* $k = 1, 2, \ldots, \infty$, *with linear part* $T_1$. *Let* $\mu_1, \ldots \mu_s$ *be the eigenvalues of* $T_1$ *counted with their algebraic multiplicities as roots of the characteristic equation of* $T_1$, *and suppose that for* $i = 1, \ldots, s$,

$$\mu_i \neq \mu_1^{m_1} \cdots \mu_s^{m_s}$$

*whenever the* $m_j$ *are positive integers with*

$$2 \leq m_1 + \ldots + m_s \leq k .$$

*Then there is a unique* $R$ *in* $\bar{\mathcal{F}}^k$ *with linear part* $R_1 = 1$ *such that* $RTR^{-1} = T_1$.

*Proof.* Extend the ground field to $\mathbb{C}$, and suppose first that $T_1$ is diagonalizable over $\mathbb{C}$. Choose coordinates so that $T_1$ is diagonal. We want to solve the equation $T_1 R = RT$; that is, the equation

$$T_1(x + R_2x^2 + R_3x^3 + \ldots) =$$
$$(T_1x + T_2x^2 + T_3x^3 + \ldots) + R_2(T_1x + T_2x^2 + T_3x^3 + \ldots)^2 + \ldots ,$$

where $Tx = \sum_j T_j x^j$ and $Rx = \sum_j R_j x^j$. Comparing coefficients of $x^j$

we see that we must have

$$T_1 R_j x^j = R_j (T_1 x)^j + \ell.\text{o.t.} \,, \tag{2}$$

where $\ell$.o.t. stands for lower order terms; that is, terms involving $R_i$ with $i < j$ , which we take to be already uniquely determined by induction. Explicitly, if we let

$$(R_j x^j)_i = \sum_{m_1 + \ldots + m_s = j} r_{i,m_1 \ldots m_s} x_1^{m_1} \ldots x_s^{m_s} \,,$$

this equation is

$$\mu_i r_{i,m_1 \ldots m_s} = r_{i,m_1 \ldots m_s} \mu_1^{m_1} \ldots \mu_s^{m_s} + \ell.\text{o.t.} \,,$$

and this has a unique solution by hypothesis. Since R is unique it is real. (We also have $\bar{R}\bar{T}\bar{R}^{-1} = \bar{T}_1$ ; that is, $\bar{R}T\bar{R}^{-1} = T_1$ .)

The result remains true without the assumption that $T_1$ is diagonalizable. Let $L(T_1)$ be the linear transformation on $L^j_{\text{sym}}(\mathbb{R}^s, \mathbb{R}^s)$ given by

$$(L(T_1)R_j)x^j = T_1 R_j x^j - R_j (T_1 x)^j \,.$$

The above argument shows that if $T_1$ is diagonalizable the eigenvalues of $L(T_1)$ are all numbers of the form

$$\mu_i - \mu_1^{m_1} \ldots \mu_s^{m_s}$$

with $i = 1, \ldots, s$; $m_1, \ldots, m_s \geq 0$; $m_1 + \ldots + m_s = j$ . By continuity, this remains true for all $T_1$ , and $L(T_1)$ is invertible by the hypothesis of the theorem, so that (2) has a unique solution $R_j$ . QED

By $o(x^\infty)$ we mean a function which is $o(x^j)$ for all $j < \infty$ .

*Theorem* 4. *Let* $X$ *be a* $C^k$ *vector field* $k = 1, 2, \ldots, \infty$, *defined in a neighborhood of* $0$ *in* $\mathbb{R}^s$, *with* $X(0) = 0$. *Let* $\lambda_1, \ldots, \lambda_s$ *be the eigenvalues of* $DX(0)$, *and suppose that for* $i = 1, \ldots, s$,

$$\lambda_1 \neq m_1\lambda_1 + \ldots + m_s\lambda_s$$

*whenever the* $m_j$ *are positive integers with*

$$2 \leq m_1 + \ldots + m_s \leq k .$$

*Then there is a local* $C^k$ *diffeomorphism* $R$ *at* $0$ *with* $R(0) = 0$, $DR(0) = 1$ *such that*

$$(R_* X)x = DX(0)x + o(x^k) . \qquad (3)$$

*Proof.* Let $U(t)$ be the local flow generated by $X$, let $X_0$ be the linear vector field $X_0 x = DX(0)x$, and let $U_0(t)$ be the flow $e^{tX_0}$ generated by $X_0$. Each $U(t)$ and $U_0(t)$ are in $\mathcal{G}^k$; let $\tilde{U}(t)$ and $\tilde{U}_0(t)$ be their images in $\tilde{\mathcal{F}}^k$. Then $\tilde{U}_0(t)$ is the linear part of $\tilde{U}(t)$ and it is simply $e^{tX_0}$, which has eigenvalues

$$e^{t\lambda_1}, \ldots, e^{t\lambda_s} .$$

Therefore Theorem 3 applies, and for each $t$ there is an $\tilde{R}$ in $\tilde{\mathcal{F}}^k$ with $D\tilde{R}(0) = 1$ such that $\tilde{R}\tilde{U}(t)\tilde{R}^{-1} = \tilde{U}_0(t)$. An $\tilde{R}$ which works for $t$ also works for $2t$, and by the uniqueness assertion of Theorem 3, $\tilde{R}$ is independent of $t$. Since $\mathcal{G}^k \to \tilde{\mathcal{F}}^k$ is surjective, there is an $R$ in $\mathcal{G}^k$ with image $\tilde{R}$, so that $R(0) = 0$, $DR(0) = 1$, and

$$RU(t)R^{-1}x = U_0(t)x + o(x^k) .$$

Let $V(t) = RU(t)R^{-1}$ and $Y = R_*X$, so that $Y$ generates $V(t)$. To conclude the proof we need only show that if $X_0$ and $Y$ are two vector fields generating local $C^k$ flows $U_0(t)$ and $V(t)$ and if the Taylor series of $U_0(t)$ and $V(t)$ at some point $x_0$ (here $x_0 = 0$) are equal to order $k$ for all sufficiently small $t$, then the Taylor series of $X_0$ and $Y$ at $x_0$ agree to order $k$. For $k = 0$ this is clear: $U_0(t)x_0 = V(t)x_0$ for $t$ sufficiently small, so $X_0x_0 = Yx_0$. Suppose it is true for $k-1$, and consider the first variational equation. Then the Taylor series of $DU_0(t)$ and $DV(t)$ at $x_0$ are equal to order $k-1$, so the Taylor series of $DX_0$ and $DY$ at $x_0$ are equal to order $k-1$, and the induction is complete. QED

We will call the condition on the $\lambda$'s in Theorem 4 the <u>eigenvalue condition</u> (to order $k$); the condition on the $\mu$'s in Theorem 3 the <u>multiplicative eigenvalue condition</u>. Notice that if the eigenvalue condition holds, $DX(0)$ must be invertible. Furthermore, if $\lambda$ is an eigenvalue of $DX(0)$ then $-\lambda$ cannot be. (Thus Hamiltonian vector fields never satisfy the eigenvalue condition.) Since imaginary eigenvalues occur in conjugate pairs, none of the eigenvalues can be purely imaginary.

Suppose that $X$ is a $C^\infty$ vector field with $X(0) = 0$, let $X_0$ be the linear vector field $X_0x = DX(0)x$, and suppose that $Xx = X_0x + o(x^\infty)$. It is not always true that there is local diffeomorphism $F$ such that $F_*X = X_0$. For example, let

$$X_0\binom{x}{y} = \binom{y}{-x},$$

let $\varphi: \mathbb{R} \longrightarrow \mathbb{R}$ be a $C^\infty$ function whose Taylor series at the origin is 1 but which is not identically 1 in any neighborhood of the origin, and let $X = \varphi(r)X_0$ where $r^2 = x^2 + y^2$. The orbits of $X_0$ and $X$ are circles with center the origin, but the period of each $X_0$ orbit is $2\pi$ while the periods of the $X$ orbits vary with position, being given by $2\pi/\varphi(r)$. Any intrinsically defined property of a vector field is preserved by a diffeomorphism, so the periods of the orbits of $F_*X$ cannot be constant and $F_*X$ cannot be equal to $X_0$. Notice however that the eigenvalues of $X_0$ are $i$ and $-i$, so that $X_0$ does not satisfy the eigenvalue condition. Thus there are two types of obstacles to linearizing a vector field, one arithmetic and the other analytic. The eigenvalue condition eliminates both obstacles--it remains to show this for the analytic obstacle. Notice that in the example given above, since $X_0$ has closed orbits the flow generated by $X_0$ never leads into a region where the perturbed vector field $X$ is arbitrarily close to $X_0$, so that wave operators cannot exist.

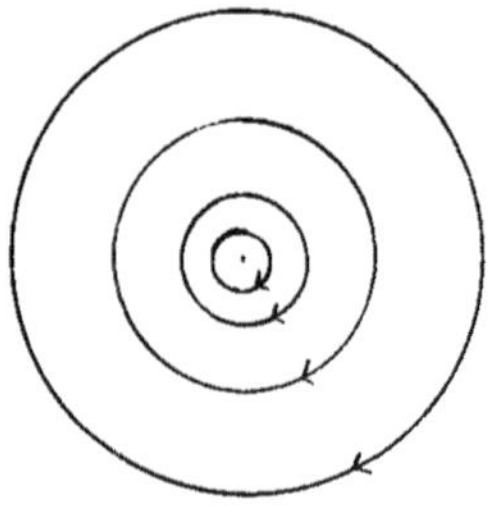

Figure 3. Orbits for $X_0\binom{x}{y} = \binom{y}{-x}$.

Theorem 5. *Let* $U$ *be an open subset of a Banach space* $E$, *let* $X: U \longrightarrow E$ *be Lipschitz with Lipschitz constant* $\kappa$, *let* $\varphi$ *be the local flow generated by* $X$, *and let* $x$ *and* $y$ *be in* $U$. *Then for positive* $t$,

$$\|\varphi(t,x) - \varphi(t,y)\| \le e^{\kappa t}\|x-y\| .$$

*Suppose that* $X = X_0 + X_1$ *where* $X_0$ *and* $X_1$ *are locally Lipschitz, and let* $\varphi_0$ *be the local flow generated by* $X_0$. *Then for positive* $t$,

$$(4) \qquad \|\varphi(t,x) - \varphi_0(t,x)\| \le \int_0^t e^{\kappa(t-s)}\|X_1(\varphi_0(s,x))\|ds .$$

*Proof.* We have

$$\varphi(t,x) = x + \int_0^t X(\varphi(s,x))ds ,$$

$$\varphi(t,y) = y + \int_0^t X(\varphi(s,y))ds ,$$

so that

$$\|\varphi(t,x) - \varphi(t,y)\| \le \|x-y\| + \int_0^t \kappa\|\varphi(s,x) - \varphi(s,y)\|ds .$$

Let $f(t) = \|\varphi(t,x) - \varphi(t,y)\|$. We claim that $f$ must be smaller than the solution $F$ of the corresponding equation $F(t) = \|x-y\| + \int_0^t \kappa F(s)ds$, which may be written as the differential equation

$$F' = \kappa F , \qquad F(0) = \|x-y\| .$$

To see this, define $\Phi$ by

$$(\Phi g)(t) = \|x-y\| + \int_0^t \kappa g(s)ds .$$

Then

$$f \le \Phi f \le \Phi^2 f \le \Phi^3 f \le \dots .$$

But the limit is the solution $F$ of the equation, so $f \leq F$. Clearly, $F(t) = e^{\kappa t}\|x-y\|$.

We prove the second statement in the same way:

$$\varphi(t,x) = x + \int_0^t (X_0 + X_1)(\varphi(s,x))ds \ ,$$

$$\varphi_0(t,x) = x + \int_0^t X(\varphi_0(s,x))ds \ ,$$

$$\varphi(t,x) - \varphi_0(t,x) =$$

$$\int_0^t [(X_0 + X_1)\varphi(s,x) - (X_0 + X_1)\varphi_0(s,x)]ds + \int_0^t X_1(\varphi_0(s,x))ds \ ,$$

so that

$$\|\varphi(t,x) - \varphi_0(t,x)\| \leq \int_0^t \kappa\|\varphi(s,x) - \varphi_0(s,x)\|ds + \int_0^t \|X_1(\varphi_0(s,x))\|ds \, .$$

As before, $\|\varphi(t,x) - \varphi_0(t,x)\|$ is smaller than the solution of the corresponding equation, which is given by the right hand side of (4). QED

We are now ready to prove the Sternberg linearization theorem. First we prove a special case, although it will be included in the general case, because it is easier. We use the notation $U(t)$ for the local flow generated by a vector field $X$, so that $U(t)x = \varphi(t,x)$.

Theorem 6. *Let $X$ be a $C^\infty$ vector field defined in a neighborhood of $0$ in $\mathbb{R}^s$ with $X(0) = 0$. Let $X_0x = DX(0)x$ and suppose that each eigenvalue $\lambda$ of $X_0$ satisfies $\operatorname{Re}\lambda < 0$ and that*

$$Xx = X_0x + o(x^\infty) \, .$$

*Let $U_0(t)$ be the flow generated by $X_0$ and let $U(t)$ be the local flow generated by $X$. Then*

$$W_- x = \lim_{t \to \infty} U(-t)U_0(t)x$$

<u>exists and is a local</u> $C^\infty$ <u>diffeomorphism at</u> 0 <u>such that</u>

$$(W_-^{-1})_* X = X_0$$

<u>in a neighborhood of</u> 0 .

<u>Proof.</u> Without loss of generality we may assume that $X$ is globally defined with a global Lipschitz constant $\kappa$, so that (as is easily seen) $U(t)$ is globally defined.

Since each eigenvalue $\lambda$ of $X_0$ satisfies $\operatorname{Re}\lambda < 0$, there are constants $C < \infty$ and $c > 0$ such that

$$\|U_0(t)\| \leq Ce^{-ct}$$

for all $t \geq 0$. (This is easily seen by writing $X_0$ in Jordan canonical form, since $U_0(t) = e^{tX_0}$.)

Define $X_1$ by $X = X_0 + X_1$. We claim that

(5) $$\|U(-t)U_0(t)x - x\| \leq \int_0^t e^{\kappa s}\|X_1(U_0(s)x)\|ds .$$

To see this, let $y = U_0(t)x$, so that $x = U_0(-t)y$. By Theorem 5,

$$\|U(-t)y - U_0(-t)y\| \leq \int_0^t e^{\kappa(t-s)}\|X_1(U_0(-s)y)\|ds$$

$$= \int_0^t e^{\kappa r}\|X_1(U_0(r-t)y)\|dr = \int_0^t e^{\kappa r}\|X_1(U_0(r)x)\|dr ,$$

which proves (5).

Now let $t_1 = t_2 + t$ with $t_1$ and $t_2$ positive. Again by Theorem 5, and by (5),

(6) $$\|U(-t_1)U_0(t_1)x - U(-t_2)U_0(t_2)x\| =$$
$$\|U(-t_2)U(-t)U_0(t)U_0(t_2)x - U(-t_2)U_0(t_2)x\| \le$$
$$e^{\kappa t_2}\|U(-t)U_0(t)U_0(t_2)x - U_0(t_2)x\| \le$$
$$e^{\kappa t_2} \int_0^t e^{\kappa s}\|X_1(U_0(s+t_2)x)\|ds \ .$$

Now $X_1x = o(x^\infty)$ , so if $k > 0$ and $t_2$ is large enough,

$$\|X_1(U_0(s+t_2)x)\| \le \|U_0(s+t_2)x\|^k \le C^k e^{-ck(s+t_2)}\|x\|^k \ ,$$

so that (6) is smaller than

$$e^{\kappa t_2} \int_0^\infty e^{\kappa s} C^k e^{-ck(s+t_2)}\|x\|^k ds$$
$$= e^{\kappa t_2} C^k\|x\|^k e^{-ckt_2} \int_0^\infty e^{(\kappa - ck)s} ds$$
$$= \frac{e^{-(ck-\kappa)t_2}}{ck-\kappa} \ C^k\|x\|^k$$

if $ck > \kappa$ . But this tends to $0$ . Consequently $W_-$ exists, and for $x$ in a bounded set $U(-t)U_0(t)x$ converges uniformly. Hence $W_-$ is continuous.

Recall the first variational equation. If $\psi(t,x) = D_2\varphi(t,x)$ where $\varphi(t,x) = U(t)x$ , then the pair $\varphi,\psi$ satisfies

$$\frac{d}{dt}\begin{pmatrix}\varphi(t,x)\\ \psi(t,x)\end{pmatrix} = \begin{pmatrix}X(\varphi(t,x))\\ DX(\varphi(t,x))\cdot\psi(t,x)\end{pmatrix} = X'\begin{pmatrix}\varphi(t,x)\\ \psi(t,x)\end{pmatrix} \ ,$$

where $X'$ is defined by this equation, and similarly for $X_0'$ . Now $X'$ and $X_0'$ satisfy the same hypotheses as before: $X_0'$ is linear, has the same eigenvalues as $X_0$ (with higher multiplicities), and $X'x = X_0'x + o(x^\infty)$. The corresponding flows are

$$U'(t)\binom{x}{\xi} = \begin{pmatrix} U(t)x \\ DU(t)x\cdot\xi \end{pmatrix},$$

$$U_0'(t)\binom{x}{\xi} = \begin{pmatrix} U_0(t)x \\ DU_0(t)x\cdot\xi \end{pmatrix} = \begin{pmatrix} U_0(t)x \\ U_0(t)\xi \end{pmatrix}.$$

Therefore

$$U'(-t)U_0'(t)\binom{x}{\xi} =$$

$$\begin{pmatrix} U(-t)U_0(t)x \\ (DU(-t))(U_0(t)x)\cdot DU_0(t)x\cdot\xi \end{pmatrix} = \begin{pmatrix} U(-t)U_0(t)x \\ D(U(-t)U_0(t))x\cdot\xi \end{pmatrix}.$$

By what we have already proved, $U'(-t)U_0'(t)$ converges to a continuous mapping. Hence $D(U(-t)U_0(t))$ converges to a continuous mapping, so that $W_-$ is $C^1$. By induction, $W_-$ is $C^\infty$.

It is clear that $DW_-(0)$ is invertible; in fact, $W_-x = x + o(x^\infty)$ so that $DW_-(0) = 1$. Hence (Theorem 3, §2) $W_-$ is a local $C^\infty$ diffeomorphism. As in the proof of Theorem 1, $(W_-^{-1})_*X = X_0$ in a neighborhood of $0$. QED

Theorem 7. *Let $X$ be a $C^\infty$ vector field in a neighborhood of $0$ in $\mathbb{R}^s$ with $X(0) = 0$ such that $DX(0)$ satisfies the eigenvalue condition and each eigenvalue $\lambda$ of $DX(0)$ satisfies $\mathrm{Re}\,\lambda < 0$. Then there is a local $C^\infty$ diffeomorphism $R$ at $0$ such that $R_*X$ is linear in a neighborhood of $0$.*

<u>Proof.</u> This is an immediate consequence of Theorems 4 and 6.

We say that two $C^\infty$ functions are <u>equal to infinite order</u> at a point if they have the same Taylor series there, and that they are equal to infinite order on a set if they are equal to infinite order at each point in a set.

The next theorem is a technical lemma from which the Sternberg linearization theorem will follow easily.

<u>Theorem 8. Let</u> $X$ <u>be a</u> $C^\infty$ <u>vector field on</u> $\mathbb{R}^s$ , <u>with</u> $X(0) = 0$ , <u>such that each</u> $D^jX$ <u>satisfies a global Lipschitz condition.</u> <u>Let</u> $X_0x = DX(0)x$ , <u>let</u> $U(t)$ <u>and</u> $U_0(t)$ <u>be the flows generated by</u> $X$ <u>and</u> $X_0$ , <u>and define</u> $X_1$ <u>by</u> $X = X_0 + X_1$ . <u>Suppose there is a linear subspace</u> $N$ , <u>invariant under</u> $X_0$ , <u>and a positive integer</u> $\ell$ <u>such that for all</u> $m \geq 0$ <u>and</u> $j = 0,1,2,\ldots$ <u>there is a</u> $\delta > 0$ <u>such that if</u> $\|z-N\| \leq \delta$ <u>then</u>

$$\|D^jX_1(z)\| \leq \|z-N\|^m \|z\|^\ell .$$

<u>Let</u> $E$ <u>be the linear subspace of all</u> $x$ <u>in</u> $\mathbb{R}^s$ <u>such that</u>

$$\lim_{t \to \infty} \|U_0(t)x-N\| = 0 .$$

<u>Then for all</u> $j = 0,1,2,\ldots$ <u>and</u> $x$ <u>in</u> $E$ ,

$$D^j(U(-t)U_0(t))x$$

<u>converges as</u> $t \longrightarrow \infty$ <u>and the limit is continuous in</u> $x$ <u>for</u> $x$ <u>in</u> $E$. <u>Let</u>

$$W_-x = \lim_{t \to \infty} U(-t)U_0(t)x$$

<u>for</u> $x$ <u>in</u> $E$ . <u>Then</u> $W_-$ <u>has a</u> $C^\infty$ <u>extension</u> $G$ <u>to</u> $\mathbb{R}^s$ <u>which is</u>

<u>the identity to infinite order in a neighborhood of</u> $0$ <u>in</u> $N$ <u>and such that in a neighborhood of</u> $0$ <u>in</u> $E$, $(G^{-1})_* X = X_0$ <u>to infinite order</u>.

<u>Proof.</u> Since $N$ is invariant under $X_0$, $N \subset E$. On the quotient space $E/N$, $U_0(t) \longrightarrow 0$ as $t \longrightarrow \infty$, so there are constants $C < \infty$ and $c > 0$ such that

$$\|U_0(t)x - N\| \leq Ce^{-ct}\|x\|$$

for all $x$ in $E$ and $t \geq 0$. Let $\kappa$ be a Lipschitz constant for $X$ and $X_0$ and choose $m$ so that

$$mc > \kappa + \ell\kappa .$$

Let $K$ be a compact set in $E$ and let $t_2$ be large enough so that

$$\|U_0(s)x - N\| \leq \delta$$

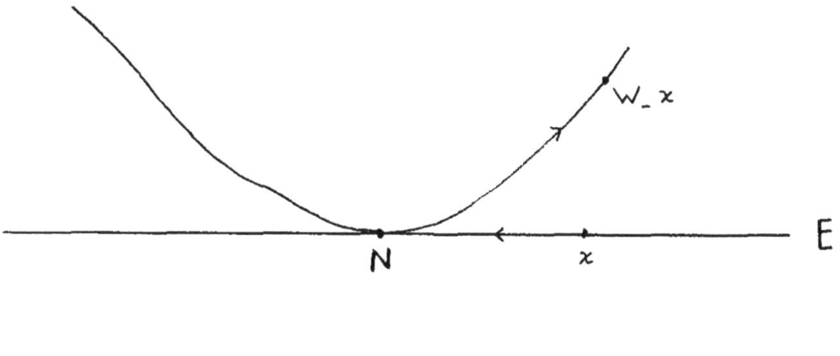


Figure 4. Construction of $W_-$ (Theorem 8).

whenever $s \geq t_2$ and $x$ is in $K$ (where $\delta$ is as in the hypothesis of the theorem). Let $t_1 \geq t_2$, $t_1 = t_2 + t$. Then by Theorem 5, as in the proof of Theorem 6, for $x$ in $K$

$$\begin{aligned}
&\|U(-t_1)U_0(t_1)x - U(-t_2)U_0(t_2)x\| = \\
&\quad \|U(-t_2)U(-t)U_0(t)U_0(t_2)x - U(-t_2)U_0(t_2)x\| \leq \\
&\quad e^{\kappa t_2}\|U(-t)U_0(t)U_0(t_2)x - U_0(t_2)x\| \leq \\
&\quad e^{\kappa t_2} \int_0^t e^{\kappa s}\|X_1(U_0(s+t_2)x)\|ds \leq \\
&\quad e^{\kappa t_2} \int_0^\infty e^{\kappa s} C^m e^{-mc(s+t_2)}\|x\|^m\|U_0(s+t_2)x\|^{\ell} ds \leq \\
&\quad e^{\kappa t_2} \int_0^\infty e^{\kappa s} C^m e^{-mc(s+t_2)}\|x\|^m e^{\kappa\ell(s+t_2)}\|x\|^{\ell} ds = \\
&\quad \frac{e^{-(mc-\kappa-\ell\kappa)t_2}\|x\|^{\ell+m}C^m}{mc-\kappa-\ell\kappa} \longrightarrow 0 .
\end{aligned}$$

Therefore $W_-$ exists and is continuous on $E$. Notice also that since $m$ may be chosen arbitrarily large, this shows that if $W_-$ is $C^\infty$ (and we shall prove this below) then it is the identity to infinite order on $N$.

Now consider the first variational equations of $X$ and $X_0$, on $\mathbb{R}^s \oplus L(\mathbb{R}^s)$. They are given by

$$X'\binom{x}{\xi} = \begin{pmatrix} X(x) \\ DX(x)\cdot\xi \end{pmatrix}, \qquad X_0'\binom{x}{\xi} = \begin{pmatrix} X_0(x) \\ DX_0(x)\cdot\xi \end{pmatrix} = \begin{pmatrix} X_0(x) \\ X_0\cdot x \end{pmatrix}.$$

We let $N'$ be the space of all $\binom{x}{\xi}$ with $x$ in $N$. We see that $X'$, $X_0'$, and $N'$ again satisfy the hypotheses of the theorem, with

$\ell$ replaced by $\ell+1$. The space $E'$ is the space of all $\binom{x}{\xi}$ with $x$ in $E$. Let $U'(t)$ and $U_0'(t)$ be the flows generated by $X'$ and $X_0'$. By what we have already shown, we know that

$$U'(-t)U_0'(t)\binom{x}{\xi}$$

converges to a continuous mapping for all $x$ in $E$, and as in the proof of Theorem 6, this implies that $D(U(-t)U_0(t))x$ converges as $t \longrightarrow \infty$ to a continuous function of $x$ in $E$, so that $W_-$ is $C^1$ on $E$. By induction, $D^j(U(-t)U_0(t))x$ converges as $t \longrightarrow \infty$ for $x$ in $E$, and $W_-$ is $C^\infty$.

Let

$$G_j(x) = \frac{1}{j!} \lim_{t \to \infty} D^j(U(-t)U_0(t))x$$

for $x$ in $E$ and $j = 0,1,2,\ldots$. Let $\alpha\colon \mathbb{R} \longrightarrow \mathbb{R}$ be as in the proof of Theorem 2 ($\alpha$ is $C^\infty$, $0 \le \alpha \le 1$, $\alpha = 1$ in a neighborhood of $0$, and $\alpha$ has support in $[-1,1]$). Let $F$ be a complementary space to $E$, $\mathbb{R}^s = E \oplus F$, and let

$$G(x \oplus y) = \sum_{j=0}^{\infty} G_j(x)y^j\alpha(\|G_j\|\|y\|^2),$$

where $\|y\|$ is the Euclidean norm. Then in a neighborhood of $0$ in $\mathbb{R}^s$, $G$ is a $C^\infty$ extension of $W_-$ which is the identity to infinite order on $N$, and

$$D^jG(x) = j!G_j(x)$$

for $x$ in $E$ and $j = 0,1,2,\ldots$. ($\|G_j\| = \sup_x \|G_j(x)\|$ for $x$ near $0$.)

Since $G$ is an extension of $W_-$,

$$U(-t)GU_0(t)x = Gx$$

for $x$ in a neighborhood of $0$ in $E$. For each fixed $s$ and for such $x$,

$$D^j(U(-s)U(-t)U_0(t)U_0(s))x \longrightarrow D^jGx .$$

But $U(-t)U_0(t) \longrightarrow G$ together with all derivatives in a neighborhood of $0$ in $E$, so that

$$D^j(U(-s)GU_0(s))x \longrightarrow D^jGx .$$

That is, $U(-s)GU_0(s) = G$ to infinite order in a neighborhood of $0$ in $E$, so that

$$U_0(s) = G^{-1}U(s)G$$

to infinite order in a neighborhood of $0$ in $E$. By the argument given at the end of the proof of Theorem 4, this implies that $X_0 = (G^{-1})_*X$ to infinite order on $E$. QED

Theorem 9 (Sternberg linearization theorem). *Let $X$ be a $C^\infty$ vector field in a neighborhood of $0$ in $\mathbb{R}^s$ with $X(0) = 0$ such that $DX(0)$ satisfies the eigenvalue condition. Then there is a $C^\infty$ local diffeomorphism $F$ at $0$ such that $F_*X$ is linear in a neighborhood of $0$.*

Proof. By Theorem 4 there is a $C^\infty$ local diffeomorphism $R$ at $0$ such that $R_*X = X_0$ to infinite order at $0$, where $X_0x = DX(0)x$. Therefore we may assume that $X = X_0$ to infinite order at $0$.

Let $E_+$ be the stable linear manifold for $X_0$; that is, the space of all vectors $x$ such that $U_0(t)x \longrightarrow 0$ as $t \longrightarrow \infty$, where

$U_0(t) = e^{tX_0}$ . Let $f$ be a $C^\infty$ function which is 1 in a neighborhood of 0 with compact support in the set where $X$ is defined, and replace $X$ by $fX$ . Then $X$, $X_0$ , and $N = 0$ satisfy the hypotheses of Theorem 8, with $\ell = 0$ and $E = E_+$ . Therefore there is a local $C^\infty$ diffeomorphism $G$ at 0 such that $G_*X = X_0$ to infinite order on a neighborhood of 0 in $E_+$ .

Let $E_-$ be the unstable linear manifold for $X_0$ ; that is, the stable linear manifold for $-X_0$ . If we apply the above result to $-X$ , we see that we may assume that $X = X_0$ to infinite order in a neighborhood $U$ of 0 in $E_-$ . Let $f$ be a $C^\infty$ function which is 1 in a neighborhood of 0 with compact support the intersection

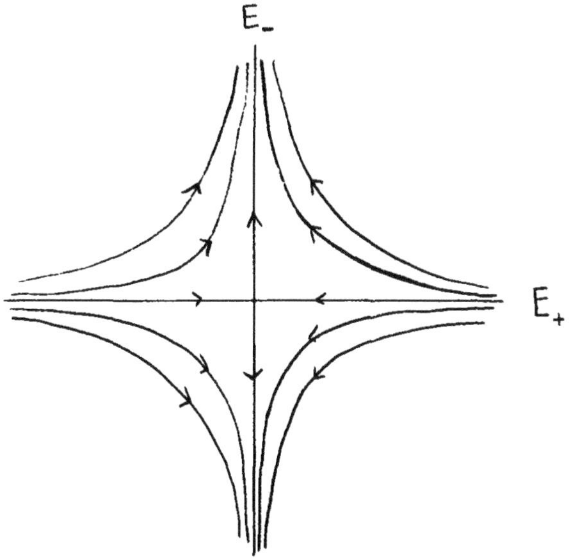


Figure 5. Linear flow near an elementary critical point (Theorem 9).

of which with $E_-$ is contained in $U$, and let $\tilde{X} = X_0 + f\cdot(X-X_0)$. Then $\tilde{X} = X_0$ outside a compact set and $\tilde{X} = X_0$ to infinite order on $E_-$. Then $\tilde{X}$, $X_0$, and $N = E_-$ satisfy the hypotheses of Theorem 8 with $\ell = 0$ and $E = \mathbb{R}^s$. Therefore there is a $C^\infty$ local diffeomorphism $F$ at $0$ such that $F_*\tilde{X} = X_0$ in a neighborhood of $0$. Since $\tilde{X} = X$ near $0$, the proof is complete.

We conclude this section by giving some complements the proofs of which are only sketched.

Let $X$ be a $C^m$ vector field defined near $0$ in $\mathbb{R}^s$, with $X(0) = 0$, such that $DX(0)$ satisfies the eigenvalue condition to order $m$. By Theorem 4 there is a $C^m$ local diffeomorphism $R$ such that $R_*X$ is linear to order $m$. Now consider the proof of Theorem 8 applied to $R_*X$ with $N = 0$, $\ell = 0$. For the constant $c$ we may take any number $c_+ < \min\{|\mathrm{Re}\,\lambda| : \mathrm{Re}\,\lambda < 0\}$ and for the Lipschitz constant $\kappa$ we may take any number $\kappa > \max|\mathrm{Re}\,\lambda|$ (after modifying $R_*X$ away from the origin). Then $U(-t)U_0(t)$ converges on $E_+$ provided $mc_+ > \kappa$, and $D^j(U(-t)U_0(t))$ converges on $E_+$ provided $mc_+ > (j+1)\kappa$, since $\ell$ is replaced by $\ell+1$ at each step of the induction. If $j$ satisfies this inequality, there is a local $C^j$ diffeomorphism $G$ such that $(G^{-1})_*R_*X$ is linear to order $j$ on $E_+$. Similarly, if $jc_- > (k+1)\kappa$, where $c_- < \min\{|\mathrm{Re}\,\lambda| : \mathrm{Re}\,\lambda > 0\}$, there is a $C^k$ local diffeomorphism $H$ such that $(H^{-1})_*(G^{-1})_*R_*X$ is linear near $0$. In this way we obtain the following theorem:

Theorem 10. Let $m$ and $k$ be integers $\geq 1$, let $X$ be a $C^m$ vector field in a neighborhood of $0$ in $\mathbb{R}^s$ with $X(0) = 0$ such *that* DX(0) *satisfies the eigenvalue condition to order* m, *and suppose that*

$$k < \frac{\min\{|\mathrm{Re}\ \lambda| : \mathrm{Re}\ \lambda > 0\}}{\max|\mathrm{Re}\ \lambda|}\left(\frac{\min\{|\mathrm{Re}\ \lambda| : \mathrm{Re}\ \lambda < 0\}}{\max|\mathrm{Re}\ \lambda|}\ m-1\right) -1 .$$

*Then there is a local* $C^k$ *diffeomorphism* F *at* 0 *such that* $F_*X$ *is linear in a neighborhood of* 0 .

The Sternberg linearization theorem may be formulated and proved in an entirely analogous manner for mappings instead of vector fields:

*Theorem* 11 (*Sternberg linearization theorem for mappings*). *Let* T *be a* $C^\infty$ *mapping defined in a neighborhood of* 0 *in* $\mathbb{R}^s$ *with* T(0) = 0 *such that* DT(0) *satisfies the multiplicative eigenvalue condition. Then there is a* $C^\infty$ *local diffeomorphism* F *at* 0 *such that* $FTF^{-1}$ *is linear in a neighborhood of* 0 .

The proof is simpler in some inessential respects than the proof for vector fields. Theorem 4 may be omitted, and Theorem 5 is replaced by the following:

*Theorem* 12. *Let* T *be Lipschitz with Lipschitz constant* K . *Then*

$$\|T^n x - T^n y\| \le K^n \|x-y\| .$$

*If* $T = T_0 + T_1$ *then*

$$\|T^n x - T_0^n x\| \le \sum_{k=1}^{n} K^{n-k} \|T_1 T_0^{k-1} x\| .$$

<u>Proof.</u> The first inequality is trivial. To prove the second inequality, write $T^n - T_0^n$ as the telescoping sum

$$T^n - T_0^n = (T^{n-1}T - T^{n-1}T_0) + (T^{n-2}TT_0 - T^{n-2}T_0T_0) + \dots + (TT_0^{n-1} - T_0T_0^{n-1}) .$$

This identity makes the second inequality obvious. QED

From a topological point of view, the only invariant of a vector field at an elementary critical point (<u>elementary</u> means that no eigenvalue of the derivative is purely imaginary) is the index $\dim E_+ - \dim E_-$ . We may use wave operators to give a simple proof of this fact in the special case that $\operatorname{Re}\lambda < 0$ for every eigenvalue. For the general case, see [7, p.244].

<u>Theorem</u> 12. <u>Let</u> $X$ <u>be</u> $C^k$ , $k = 1,2,\dots,\infty$ , <u>in a neighborhood of</u> $0$ <u>in</u> $\mathbb{R}^s$ <u>with</u> $X(0) = 0$ , <u>and suppose that every eigenvalue</u> $\lambda$ <u>of</u> $DX(0)$ <u>satisfies</u> $\operatorname{Re}\lambda < 0$ . <u>Then there is a local homeomorphism</u> $W$ <u>in a neighborhood of</u> $0$ , $W(0) = 0$ , <u>which is</u> $C^k$ <u>in a deleted neighborhood of</u> $0$ <u>and is such that</u> $W_*X = -1$ <u>in a deleted neighborhood of</u> $0$ .

<u>Proof.</u> Let $X_0x = DX(0)x$ . Since each eigenvalue $\lambda$ of $X_0$ satisfies $\operatorname{Re}\lambda < 0$ , we can give $\mathbb{R}^s$ an inner product so that for some $c > 0$ ,

$$(x,X_0x) \le -c(x,x) , \qquad x \in \mathbb{R}^s .$$

(For diagonalizable $X_0$ this is clear, and for the general case we may consider the Jordan canonical form with the 1's replaced by $\varepsilon$'s.)

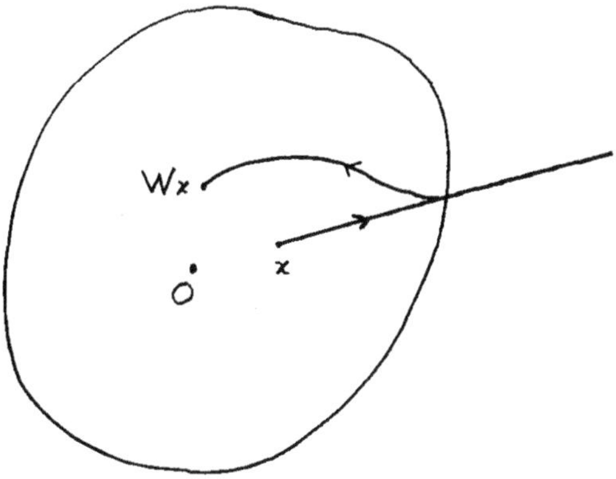


Figure 6. Construction of W (Theorem 12).

Choose a $C^\infty$ function $f$ with $0 \le f \le 1$ which is 1 in a neighborhood of 0 and has compact support in the domain of definition of $X$ and let $\tilde{X} = -1 + f\cdot(X+1)$. Then $\tilde{X} = X$ near 0, $\tilde{X} = -1$ outside a compact set, and

$$(x,\tilde{X}x) \le -c(x,x)\,, \qquad x \in \mathbb{R}^s\,,$$

for some $c > 0$ if we choose $f$ appropriately, since $Xx = X_0x + o(x)$.

Let $\tilde{U}(t)$ and $U_0(t)$ be the flows generated by $\tilde{X}$ and $-1$. Consider

$$Wx = \lim_{t\to\infty} \tilde{U}(t)U_0(-t)x\,.$$

This limit clearly exists and is $C^k$ on $\mathbb{R}^s - \{0\}$, since $U(t)U_0(-t)x$ is eventually constant in $t$. Also, $W0 = 0$, and

$$W^{-1}x = \lim_{t\to\infty} U_0(t)\tilde{U}(-t)x$$

has the same properties, since by the construction of $\tilde{X}$, $\tilde{U}(-t)x$ (for $x \neq 0$) eventually enters the region where $\tilde{X} = -1$. It is also clear that $W$ and $W^{-1}$ are continuous at $0$, so that $W$ is a homeomorphism of $\mathbb{R}^s$ onto $\mathbb{R}^s$. We have $W_*\tilde{X} = -1$ on $\mathbb{R}^s - \{0\}$, and since $\tilde{X} = X$ near $0$ the proof is complete.

This topological classification is too crude to be of much interest. See the examples of vector fields in [8, pp 372-375].

## 4. Sums and Lie products of vector fields

If $X$ is a locally Lipschitz vector field in an open subset of a Banach space, we denote by $U_X(t)$ the local flow it generates.

Theorem 1. *Let* $X$ *and* $Y$ *be locally Lipschitz vector fields defined in an open subset* $U$ *of the Banach space* $E$. *For all* $x_0$ *in* $U$ *there is a neighborhood* $V$ *of* $x_0$ *with* $V$ *contained in* $U$ *and an* $\varepsilon > 0$ *such that*

$$U_{X+Y}(t)x = \lim_{n \to \infty} \left(U_X\left(\frac{t}{n}\right)U_Y\left(\frac{t}{n}\right)\right)^n x$$

*uniformly for* $x$ *in* $V$ *and* $|t| \leq \varepsilon$.

Proof. We have

$$U_X(h) = 1 + hX + o(h),$$

and for each $x_0$ this holds uniformly for $x$ in a neighborhood of $x_0$, and similarly for $U_Y(h)$ and $U_{X+Y}(h)$. Therefore

$$U_{X+Y}(h) = U_X(h)U_Y(h) + o(h)$$

uniformly in a neighborhood of $x_0$. If we write

$$(U_{X+Y}(\frac{t}{n}))^n - (U_X(\frac{t}{n})U_Y(\frac{t}{n}))^n$$

as a telescoping sum (as in the proof of Theorem 12, §3) we see that if $\kappa$ is a Lipschitz constant for $X+Y$ near $x_0$ then

$$\|U_{X+Y}(t)x - (U_X(\frac{t}{n})U_Y(\frac{t}{n}))^n x\| \leq e^{\kappa t} n\, o(\frac{t}{n}) \longrightarrow 0$$

uniformly for $x$ in a neighborhood of $x_0$. QED

Let $X$ and $Y$ be $C^k$ vector fields, with $k \geq 1$, defined in an open set $U$ in the Banach space $E$. We define the <u>Lie product</u> $[X,Y]$ by

$$[X,Y](x) = DY(x)\cdot X(x) - DX(x)\cdot Y(x), \qquad x \in U.$$

This is a $C^{k-1}$ vector field. Recall that we are abusing the term "vector field", and it is necessary to show that if $R$ is a $C^j$ diffeomorphism, with $j \geq 2$, then

(1) $$[R_*X, R_*Y] = R_*[X,Y].$$

This is a consequence of the symmetry of the second derivative, as we now show. We have

$$R_*X = DR\circ R^{-1}\cdot X\circ R^{-1} = (DR\cdot X)\circ R^{-1}.$$

Let $Rx = y$. Then

$$D(R_*X)(y) = D(DR\cdot X)(x)\cdot DR^{-1}(y)$$

by the chain rule, and similarly for $Y$. Therefore

$$\begin{aligned}&[R_*X,R_*Y](y)\\&\quad = D(DR\cdot Y)(x)\cdot DR^{-1}(y)DR(x)X(x) - D(DR\cdot X)(x)\cdot DR^{-1}(y)DR(x)Y(x)\\&\quad = D(DR\cdot Y)(x)\cdot X(x) - D(DR\cdot X)(x)\cdot Y(x) .\end{aligned}$$

But

$$\begin{aligned}&D(DR\cdot Y)\cdot X - D(DR\cdot X)\cdot Y\\&\qquad = D^2R\cdot Y\cdot X + DR\cdot DY\cdot X - D^2R\cdot X\cdot Y - DR\cdot DX\cdot Y\\&\qquad = DR\cdot (DY\cdot X - DX\cdot Y) = DR\cdot [X,Y]\end{aligned}$$

so that

$$[R_*X,R_*Y](y) = DR(x)\cdot [X,Y](x) = R_*[X,Y](y) .$$

Theorem 2. *Let* $X$ *and* $Y$ *be* $C^2$ *vector fields defined in an open set* $U$ *of a Banach space* $E$ . *For all* $x_0$ *in* $U$ *there is a neighborhood* $V$ *of* $x_0$ *with* $V$ *contained in* $U$ *and an* $\varepsilon > 0$ *such that*

$$U_{[X,Y]}(t)x = \lim_{n\to\infty} \left(U_Y\left(-\sqrt{\tfrac{t}{n}}\right)U_X\left(-\sqrt{\tfrac{t}{n}}\right)U_Y\left(\sqrt{\tfrac{t}{n}}\right)U_X\left(\sqrt{\tfrac{t}{n}}\right)\right)^n x$$

*uniformly for* $x$ *in* $V$ *and* $0 \le t \le \varepsilon$ .

*Proof.* Since $X$ is $C^2$ so is the local flow (by Theorem 4, §2), and we have

$$\frac{d}{dh} U_X(h) = X\circ U_X(h) ,$$

$$\frac{d^2}{dh^2} U_X(h) = DX\circ U_X(h)\cdot X\circ U_X(h) ,$$

so that by Taylor's formula

$$U_X(h) = 1 + hX + \frac{h^2}{2} DX\cdot X + o(h^2)$$

uniformly for $x$ in a neighborhood of $x_0$, and similarly for $Y$. Therefore

$$U_Y(-h)U_X(-h)U_Y(h)U_X(h) =$$

$$(1-hY + \frac{h^2}{2} DY\cdot Y)\circ(1-hX + \frac{h^2}{2} DX\cdot X)\circ(1+hY + \frac{h^2}{2} DY\cdot Y)\circ(1+hX + \frac{h^2}{2} DX\cdot X) + o(h^2).$$

When we expand this we must not forget terms like $h^2 DY\cdot X$ in

$$hY\circ(1 + hX) = hY + h^2 DY\cdot X + o(h^2) .$$

We obtain

$$1 - hY + \frac{h^2}{2} DY\cdot Y + h^2 DY\cdot X - h^2 DY\cdot Y - h^2 DY\cdot X - hX + \frac{h^2}{2} DX\cdot X$$

$$- h^2 DX\cdot Y - h^2 DX\cdot X + hY + \frac{h^2}{2} DY\cdot Y + h^2 DY\cdot X + hX + \frac{h^2}{2} DX\cdot X + o(h^2)$$

$$= 1 + h^2[X,Y] + o(h^2) .$$

Therefore

$$U_{[X,Y]}(\tfrac{t}{n}) = U_Y(-\sqrt{\tfrac{t}{n}})U_X(-\sqrt{\tfrac{t}{n}})U_Y(\sqrt{\tfrac{t}{n}})U_X(\sqrt{\tfrac{t}{n}}) + o(\tfrac{t}{n})$$

uniformly in a neighborhood of $x_0$. When we write the difference of the n'th powers as a telescoping sum, as in the proof of Theorem 1, we obtain the desired result. QED

Theorem 2 shows that the Lie product has an invariant meaning, independent of the choice of coordinates, so that (1) also follows from Theorem 2.

## 5. Self-adjoint operators on Hilbert space

We shall develop briefly those aspects of Hilbert space theory which are of greatest relevance to dynamics. A knowledge of integration theory is assumed.

Let $\mathcal{H}$ be a complex vector space. A sesquilinear form on $\mathcal{H}$ is a mapping $\mathcal{H} \times \mathcal{H} \longrightarrow \mathbf{C}$ which takes each ordered pair $(u,v)$ into a complex number $\langle u,v\rangle$, such that $\langle u,v\rangle$ is conjugate linear in $u$ and linear in $v$ (we follow the physicists' convention). For a sesquilinear form, computation shows that the polarization identity holds:

$$4\langle u,v\rangle = \langle u+v,u+v\rangle - \langle u-v,u-v\rangle - i\langle u+iv,u+iv\rangle + i\langle u-iv,u-iv\rangle .$$

This means that any sesquilinear form is determined by the associated quadratic form $u \rightsquigarrow \langle u,u\rangle$. A sesquilinear form is called Hermitean in case $\langle v,u\rangle = \overline{\langle u,v\rangle}$, so that a sesquilinear form is Hermitean if and only if $\langle u,u\rangle$ is always real. The form is called positive in case $\langle u,u\rangle \geq 0$ (so that a positive form is necessarily Hermitean), and strictly positive in case $\langle u,u\rangle > 0$ unless $u = 0$.

For a positive form we have the Schwarz inequality:

$$(1) \qquad |\langle u,v\rangle| \leq \langle u,u\rangle^{\frac{1}{2}}\langle v,v\rangle^{\frac{1}{2}} .$$

To prove this, observe that

$$0 \leq \langle u+v,u+v\rangle = \langle u,u\rangle + \langle v,v\rangle + \langle u,v\rangle + \langle v,u\rangle ,$$

so that

$$-2\mathrm{Re}\langle u,v\rangle \leq \langle u,u\rangle + \langle v,v\rangle .$$

We may multiply $v$ by a number of absolute value $1$ to ensure that $-\mathrm{Re}\langle u,v\rangle = |\langle u,v\rangle|$ , so that

$$2|\langle u,v\rangle| \leq \langle u,u\rangle + \langle v,v\rangle . \tag{2}$$

Suppose $\langle v,v\rangle = 0$ . Then the left hand side of (2) is homogeneous of degree $1$ in $u$ while the right hand side is homogeneous of degree $2$ in $u$ , so both are $0$ , and similarly if $\langle u,u\rangle = 0$ . If neither is $0$ , we may take them both to be $1$ , and (2) implies (1).

For a strictly positive form we define

$$\|u\| = \langle u,u\rangle^{\frac{1}{2}} .$$

The *triangle inequality*

$$\|u+v\| \leq \|u\| + \|v\|$$

holds, so $u \rightsquigarrow \|u\|$ is a norm.

A *Hilbert space* $\mathcal{H}$ is a complex vector space with a strictly positive sesquilinear form which is complete in the associated norm. (Thus a Hilbert space is a Banach space together with a sesquilinear form such that $\|u\|^2 = \langle u,u\rangle$ for all $u$ .)

If $\mathcal{M}$ is any subset of the Hilbert space $\mathcal{H}$ , we define $\mathcal{M}^{\perp}$ to be the set of all $u$ in $\mathcal{H}$ such that $\langle u,v\rangle = 0$ for all $v$ in $\mathcal{M}$ . This is clearly a closed linear subspace of $\mathcal{H}$ .

*Theorem* 1 (*projection theorem*). *Let* $\mathcal{M}$ *be a closed linear subspace of the Hilbert space* $\mathcal{H}$ . *Then* $\mathcal{H} = \mathcal{M} \oplus \mathcal{M}^{\perp}$ *and* $\mathcal{M} = \mathcal{M}^{\perp\perp}$ .

*Proof.* Clearly $\mathcal{M} \cap \mathcal{M}^{\perp} = 0$ . Let $u$ be in $\mathcal{H}$ and let

$$d = \inf_{x \in \mathcal{M}} \|u-x\| .$$

Let $x_n$ be a sequence in $\mathcal{M}$ such that $\|u-x_n\| \longrightarrow d$. We claim that $x_n$ is a Cauchy sequence. Computation shows that the parallelogram law

$$\|x_n-x_m\|^2 = 2\|x_n-u\|^2 + 2\|x_m-u\|^2 - 4\left\|\frac{x_n+x_m}{2} - u\right\|^2$$

holds. But $(x_n+x_m)/2$ is in $\mathcal{M}$, so that

$$\left\|\frac{x_n+x_m}{2} - u\right\|^2 \geq d^2$$

and

$$\|x_n-x_m\|^2 \leq 2\|x_n-u\|^2 + 2\|x_m-u\|^2 - 4d^2 \longrightarrow 0 .$$

Since $\mathcal{H}$ is complete, $x_n$ has a limit $x$ which is in $\mathcal{M}$ since $\mathcal{M}$ is closed, and $\|x-u\| = d$.

Since $\langle u-x+ty, u-x+ty\rangle$ for $y$ in $\mathcal{M}$ and $t$ real has a minimum at $t = 0$, its derivative there is $0$, so that $\langle y,u-x\rangle + \langle u-x,y\rangle = 2 \operatorname{Re}\langle y,u-x\rangle = 0$. This remains true if we multiply $y$ by a complex number, and so $\langle y,u-x\rangle = 0$. That is, $u-x$ is in $\mathcal{M}^{\perp}$. Thus $\mathcal{H} = \mathcal{M} \oplus \mathcal{M}^{\perp}$. By the same result applied to $\mathcal{M}^{\perp}$, $\mathcal{H} = \mathcal{M}^{\perp\perp} \oplus \mathcal{M}^{\perp}$. Since $\mathcal{M} \subset \mathcal{M}^{\perp\perp}$, we have $\mathcal{M} = \mathcal{M}^{\perp\perp}$. QED

A corollary is the Riesz representation theorem: If $v \rightsquigarrow f(v)$ is a continuous linear functional on $\mathcal{H}$ then there is a unique vector $u$ in $\mathcal{H}$ such that $\langle u,v\rangle = f(v)$ for all $v$ in $\mathcal{H}$. (Apply the projection theorem to the null space of $f$.)

If $A$ is in $L(\mathcal{H})$ then for each $u$, $v \rightsquigarrow \langle u,Av\rangle$ is a continuous linear functional, so by the Riesz representation theorem there is a unique vector, which is denoted by $A^*u$, such that

$\langle u, Av \rangle = \langle A^* u, v \rangle$ for all $v$. Clearly $A^*$ is in $L(\mathcal{H})$. It is called the <u>adjoint</u> of $A$. An operator $A$ in $L(\mathcal{H})$ is called <u>self-adjoint</u> in case $A = A^*$, <u>skew-adjoint</u> in case $A = -A^*$, and <u>normal</u> in case $AA^* = A^*A$.

If $\mathcal{M}$ is a closed linear subspace of $\mathcal{H}$, the mapping $E: u \rightsquigarrow x$ where $u = x+y$ with $x$ in $\mathcal{M}$ and $y$ in $\mathcal{M}^{\perp}$ is easily seen to be in $L(\mathcal{H})$, and $E = E^* = E^2$. It is called the <u>projection</u> onto $\mathcal{M}$. Conversely, an operator $E$ in $L(\mathcal{H})$ with $E = E^* = E^2$ is the projection onto $\mathcal{M}$ where $\mathcal{M}$ is the range of $E$.

An operator $U$ in $L(\mathcal{H})$ is <u>unitary</u> in case $U$ is bijective and $\langle Uu, Uv \rangle = \langle u, v \rangle$ for all $u$ and $v$. This is the same as saying that $U$ is invertible and $U^* = U^{-1}$. A <u>strongly continuous one-parameter unitary group</u> is a family of unitary operators $U(t)$ on $\mathcal{H}$, defined for all real $t$, such that

$$U(t+s) = U(t)U(s)$$

and such that for all $u$ in $\mathcal{H}$, $t \rightsquigarrow U(t)u$ is continuous. It is clear that $U(0) = 1$, $U(-t) = U(t)^* = U(t)^{-1}$, and that the $U(t)$ commute.

If $A$ is a self-adjoint operator in $L(\mathcal{H})$ then $U(t) = e^{itA}$ (defined by the power series expansion) is clearly invertible with $U(-t) = U(t)^* = U(t)^{-1}$, and it is a one-parameter unitary group with the stronger continuity property that $t \rightsquigarrow U(t)$ is continuous from $\mathbb{R}$ to $L(\mathcal{H})$ (norm continuity or uniform continuity).

Three topologies on $L(\mathcal{H})$ are especially useful: the <u>norm topology</u> in which a basic set of neighborhoods of $0$ is given by

$$N_{\varepsilon} = \{A \in L(\mathcal{H}): \|A\| < \varepsilon\}$$

where $\varepsilon > 0$ , the strong topology with

$$N_{\varepsilon,u_1,\ldots,u_n} = \{A \in L(\mathcal{H}): \|Au_1\| < \varepsilon,\ldots,\|Au_n\| < \varepsilon\}$$

where $\varepsilon > 0$ and $u_1,\ldots,u_n$ are in $\mathcal{H}$ , and the weak topology with

$$N_{\varepsilon,u_1,\ldots,u_n, v_1,\ldots,v_n}$$
$$= \{A \in L(\mathcal{H}): |(u_1,Av_1)| < \varepsilon,\ldots,|(u_n,Av_n)| < \varepsilon\}$$

where $\varepsilon > 0$ and $u_1,\ldots,v_n$ are in $\mathcal{H}$ . The norm topology (also called the uniform topology) is stronger than the strong topology, which is stronger than the weak topology.

We will be concerned mainly with unbounded operators. An operator $A$ on $\mathcal{H}$ is a linear transformation from a linear subspace $\mathcal{D}(A)$ , called the domain of $A$ , to $\mathcal{H}$ . Its range is denoted by $\mathcal{R}(A)$ . It is convenient to introduce a more general notion: a graph on $\mathcal{H}$ is a linear subspace $A$ of $\mathcal{H} \oplus \mathcal{H}$ . We may identify an operator with its graph, the set of all vectors of the form $u \oplus Au$ , $u \in \mathcal{D}(A)$ , in $\mathcal{H} \oplus \mathcal{H}$ . If $A$ is a graph its domain $\mathcal{D}(A)$ is the set of all $u$ in $\mathcal{H}$ such that $u \oplus v$ is in $A$ for some $v$ , and its range $\mathcal{R}(A)$ is the set of all $v$ in $\mathcal{H}$ such that $u \oplus v$ is in $A$ for some $u$ . A graph $A$ is an operator if and only if $0 \oplus v$ in $A$ implies that $v = 0$ . An operator $A$ is called closed in case its graph is closed. Thus an operator $A$ is closed if and only if $u_n \in \mathcal{D}(A)$ , $u_n \to u$ , and $Au_n \to v$ implies that $u \in \mathcal{D}(A)$

and $Au = v$. An operator is *pre-closed* if its closure (as a graph) is an operator, called the *closure* of $A$ and denoted by $\overline{A}$. An operator or graph $A$ is called *densely defined* in case $\mathcal{D}(A)$ is dense in $\mathcal{H}$. We define the operators $\rho$ and $\tau$ on $\mathcal{H} \oplus \mathcal{H}$ by

$$\rho(u \oplus v) = v \oplus u ,$$

$$\tau(u \oplus v) = (-v) \oplus u .$$

If $A$ is any graph, its *inverse* $A^{-1}$ is defined to be $\rho A$. If $A$ is any graph, its *adjoint* $A^*$ is defined to be $(\tau A)^{\perp}$.

*Theorem* 2. *Let* $A$ *be a graph on the Hilbert space* $\mathcal{H}$. *Then*

(i) $A^{*-1} = A^{-1*}$

(ii) $\overline{A}^{-1} = \overline{A^{-1}}$

(iii) $\overline{A}^{*} = \overline{A^{*}} = A^{*}$

(iv) $A = (A^{-1})^{-1}$

(v) $\overline{A} = A^{**}$

(vi) $A^*$ *is an operator if and only if* $A$ *is densely defined,*

(vii) $A^*$ *is densely defined if and only if* $\overline{A}$ *is an operator.*

*If* $A$ *is a densely defined operator,* $u$ *is in* $\mathcal{D}(A^*)$ *if and only if* $v \rightsquigarrow \langle u,Av\rangle$ *is a continuous linear functional on* $\mathcal{D}(A)$, *in which case* $A^*u$ *is the unique vector such that* $\langle A^*u,v\rangle = \langle u,Av\rangle$ *for all* $v$ *in* $\mathcal{D}(A)$. *If* $A$ *is a densely defined closed operator, so is* $A^*$, *and* $A^{**} = A$.

*Proof.* The statements (i)--(v) are trivial to verify. To see (vi), notice that $A^*$ is an operator if and only if $A^* \cap (0 \oplus \mathcal{H}) = 0$, if and only if $A^{\perp} \cap (\mathcal{H} \oplus 0) = 0$ (since $\tau$ is unitary with $\tau^2 = -1$), if and only if $\mathcal{D}(A)^{\perp} = 0$, if and only if $A$ is densely defined, by the projection theorem. By the projection theorem again, $\bar{A} = A^{\perp\perp} = A^{**}$, but by (vi), $A^{**}$ is an operator if and only if $A^*$ is densely defined, which proves (vii). The remaining statements are trivial to verify. QED

An operator $A$ is called self-adjoint in case $A = A^*$. An operator $A$ is called Hermitean (or symmetric) in case it is densely defined and $A \subset A^*$. Thus a self-adjoint operator is Hermitean, but the converse is false in general.

In our proof of the spectral theorem we will use the Riesz-Fischer theorem, which asserts that $L^2$ of a measure space is complete and hence a Hilbert space, and the other Riesz representation theorem which says that if $I$ is an interval, $C(I)$ the Banach space of continuous functions on $I$ in the supremum norm, and $\mu$ is a positive linear functional on $C(I)$, then there is a measure on $I$, also denoted by $\mu$, such that

$$\mu(f) = \int_I f(x)d\mu(x)$$

for all $f$ in $C(I)$. If $f$ is a measurable function on a measure space, the corresponding multiplication operator is the operator $M_f$: $g \rightsquigarrow fg$ on the domain $\mathcal{D}(M_f)$ of all $g$ in $L^2$ such that $fg$ is in $L^2$.

A linear operator from one Hilbert space to another is called unitary in case it is bijective and preserves inner products.

Theorem 3 (spectral theorem). *An operator* $A$ *on a Hilbert space* $\mathcal{H}$ *is self-adjoint if and only if there is a unitary operator* $\mathcal{F}$ *from* $\mathcal{H}$ *to* $L^2(M,\mu)$ *, for some measure space* $(M,\mu)$ *, such that* $\mathcal{F} A \mathcal{F}^{-1}$ *is a multiplication operator by a real measurable function.*

Proof. Suppose to begin with that $A$ is a self-adjoint operator in $L(\mathcal{H})$ . Let

$$I = [-\|A\|,\|A\|]$$

and let $p$ be a polynomial. We claim that

$$\|p(A)\| \leq \sup_{\lambda \in I} |p(\lambda)| . \tag{3}$$

To see this, let $n$ be the degree of $p$ , let $u$ be in $\mathcal{H}$ , and let $E$ be the projection onto the subspace $\mathcal{M}$ generated by $u, Au, \ldots, A^n u$ . Then $p(EAE)u = p(A)u$ . But $EAE$ is a self-adjoint operator on the finite dimensional Hilbert space $\mathcal{M}$ and so $\mathcal{M}$ has a basis of eigenvectors. (Proof: $\det(\lambda - EAE)$ is a polynomial and so has a root. Thus there is one eigenvector $e_1$ . Since $EAE$ is self-adjoint it leaves the orthogonal complement of $e_1$ in $\mathcal{M}$ invariant, so there is another eigenvector $e_2$ , and by induction there is a basis of eigenvectors.) Since $EAE$ is self-adjoint, each eigenvalue is real, and has absolute value at most $\|EAE\| \leq \|A\|$ , and so lies in $I$ . Hence $\|p(EAE)\| \leq \sup\{|p(\lambda)| : \lambda \in I\}$ . Therefore $\|p(A)u\| = \|p(EAE)u\| \leq \sup\{|p(\lambda)| : \lambda \in I\}$ . Since this is true for each $u$ , (3) holds.

Thus the mapping $p \rightsquigarrow p(A)$ from polynomials on $I$ to operators on $L(\mathcal{H})$ is continuous. But the Weierstrass theorem asserts that any continuous function $f$ on $I$ may be approximated uniformly by a polynomial. (Proof: Extend $f$ to be continuous on $\mathbb{R}$ with compact support. Then

$$\frac{1}{\sqrt{4\pi t}} \int_{-\infty}^{\infty} e^{-\frac{(\lambda-\mu)^2}{4t}} f(\mu)d\mu$$

approximates $f$ uniformly as $t \longrightarrow 0$. But it is an entire function of $\lambda$ and so the truncated Taylor series approximate it uniformly on compact sets.) Hence there is a unique continuous mapping $f \rightsquigarrow f(A)$ from $C(I)$ to $L(\mathcal{H})$ such that for $f$ a polynomial, $f(A)$ is the polynomial applied to $A$. The mapping is a homomorphism and $\bar{f}(A) = f(A)^*$, since these properties hold for polynomials.

Let $u$ be in $\mathcal{H}$ and consider the linear functional $\mu_u$ on $C(I)$ given by

$$\mu_u(f) = \langle u, f(A)u\rangle .$$

If $f \geq 0$ then $f = g^2$, with $g = \bar{g}$ in $C(I)$, so that

$$\mu_u(f) = \langle u, g^2(A)u\rangle = \langle g(A)^*u, g(A)u\rangle = \langle g(A)u, g(A)u\rangle \geq 0 .$$

Thus $\mu_u$ is a positive linear functional and there is a unique measure $\mu_u$ on $I$ such that

$$\langle u, f(A)u\rangle = \int_I f(\lambda)d\mu_u(\lambda)$$

for all $f$ in $C(I)$.

Let $\mathcal{H}_u$ be the closed linear subspace generated by $u, Au, A^2u, \ldots$ . Then $A$ restricted to $\mathcal{H}_u$ is a self-adjoint operator on $\mathcal{H}_u$ . We claim that there is a unique unitary operator

$$\mathcal{F}_u : \mathcal{H}_u \longrightarrow L^2(\mu_u)$$

such that $\mathcal{F}_u u = 1$ and $\mathcal{F}_u A \mathcal{F}_u^{-1} = M_\lambda$ , where $M_\lambda$ is the multiplication operator $(M_\lambda f)(\lambda) = \lambda f(\lambda)$ . To see this, let

$$\mathcal{F}_u(f(A)u) = f$$

for all $f$ in $C(I)$ . If $f(A)u = 0$ then $|f|^2(A)u = 0$ , so that

$$0 = \langle f(A)u, f(A)u\rangle = \int_I |f(\lambda)|^2 d\mu_u(\lambda)$$

and $f = 0$ almost everywhere with respect to the measure $\mu_u$ . Thus $\mathcal{F}_u$ is well defined from a dense set in $\mathcal{H}_u$ to a dense set in $L^2(\mu_u)$ . We have

$$\langle g(A)u, f(A)u\rangle = \langle u, (\bar{g}f)(A)u\rangle = \int_I \bar{g}(\lambda)f(\lambda)d\mu_u(\lambda)$$

and so $\mathcal{F}_u$ has a unique unitary extension $\mathcal{F}_u : \mathcal{H}_u \longrightarrow L^2(\mu_u)$ . It is clear that $\mathcal{F}_u u = 1$ and $\mathcal{F}_u A \mathcal{F}_u^{-1} = M_\lambda$ , and uniqueness is also clear.

Now we may finish the proof of the spectral theorem for the case of a self-adjoint operator $A$ in $L(\mathcal{H})$ . By Zorn's lemma there is a family of vectors $u_\alpha$ such that $\mathcal{H}_{u_\alpha}$ and $\mathcal{H}_{u_\beta}$ are orthogonal whenever $\alpha \neq \beta$ and such that $\mathcal{H}$ is the direct sum of the $\mathcal{H}_{u_\alpha}$ . Let $(M,\mu)$ be the direct sum (disjoint union) of the measure spaces $(I,\mu_{u_\alpha})$ and define the unitary operator $\mathcal{F} : \mathcal{H} \longrightarrow L^2(M,\mu)$ by letting $\mathcal{F} = \mathcal{F}_{u_\alpha}$ on $\mathcal{H}_{u_\alpha}$ . Then $\mathcal{F}$ has the desired property that $\mathcal{F} A \mathcal{F}^{-1}$ is a multiplication operator; in fact, it is $M_\lambda$ where $\lambda$ is defined by

$$(\lambda f_\alpha)(\lambda) = \lambda f_\alpha(\lambda) .$$

Let us denote $\mathcal{F} u$ by $\hat{u}$ and $\mathcal{F} A \mathcal{F}^{-1}$ by $\hat{A}$. Let $f$ be a bounded Baire function on $I$ and define $f(A)$ by

$$f(A) = \mathcal{F}^{-1} f(\hat{A}) \mathcal{F} .$$

If the $f_j$ are uniformly bounded Baire functions, $|f_j| \leq K$, converging pointwise to $f$ then

$$f_j(\hat{A}(x))\hat{u}(x) \longrightarrow f(\hat{A}(x))\hat{u}(x)$$

for all $x$ in $M$ and

$$|f_j(\hat{A}(x))\hat{u}(x) - f(\hat{A}(x))\hat{u}(x)|^2 \leq 4K^2|\hat{u}(x)|^2 .$$

By the Lebesgue dominated convergence theorem,

$$f_j(\hat{A})u \longrightarrow f(\hat{A})u$$

in $L^2(M,\mu)$, and so $f_j(A)u \longrightarrow f(A)u$ in $\mathcal{H}$. This shows that $f \rightsquigarrow f(A)$ is well-defined, independently of the choice of $\mathcal{F} : \mathcal{H} \longrightarrow L^2(M,\mu)$. Notice that the $f(A)$ all commute. In particular if $f$ is the characteristic function $\chi_B$ of a Borel set in $I$ then

$$E_B = \chi_B(A)$$

is well-defined. Since $\chi_B = \chi_B^2 = \overline{\chi}_B$, $E_B$ is a projection, called the <u>spectral projection</u> of $A$ corresponding to the Borel set $B$.

Now let $A_1,\ldots,A_n$ be a finite set of commuting self-adjoint operators in $L(\mathcal{H})$. Let $I_i = [-\|A_i\|,\|A_i\|]$ and

$$I = \prod_{i=1}^{n} I_i .$$

Let $f$ be a finite linear combination of functions of the form

$$\chi = \chi_{B_1} \cdots \chi_{B_n}$$

where the $B_i$ are Borel sets in $I_i$. Define

$$\chi(A_1,\ldots,A_n) = \chi_{B_1}(A_1) \cdots \chi_{B_n}(A_n)$$

and define $f(A_1,\ldots,A_n)$ by linearity. Noice that the $\chi_{B_i}(A_i)$ commute, since they are strong limits of polynomials in the $A_i$. We claim that

$$\|f(A_1,\ldots,A_n)\| \leq \sup_{\lambda \in I} |f(\lambda)| .$$

This is clearly true for $f = \chi$, and since $f$ may be written as a linear combination of such $\chi$'s with zero products, it is true for the general $f$. Consequently we may extend the mapping $f \rightsquigarrow f(A_1,\ldots,A_n)$ to the uniform limits of such functions; in particular, to the continuous functions on $I$. If we repeat the discussion we gave above for a single self-adjoint operator $A$ in $L(\mathcal{H})$, we obtain the spectral theorem for an n-tuple of commuting self-adjoint operators $A_1,\ldots,A_n$ in $L(\mathcal{H})$: there is a unitary operator $\overline{\mathcal{F}} : \mathcal{H} \longrightarrow L^2(M,\mu)$, for some measure space $(M,\mu)$, such that the $\overline{\mathcal{F}} A_i \overline{\mathcal{F}}^{-1}$ are multiplication operators by real measurable functions on $M$.

In particular, this result holds for a pair of commuting self-adjoint operators $A$ and $B$ in $L(\mathcal{H})$. If $C$ is in $L(\mathcal{H})$ we may write $C$ uniquely as $C = A+iB$ with $A$ and $B$ self-adjoint in $L(\mathcal{H})$, and if $C$ is normal then $A$ and $B$ commute. Thus we have the spectral theorem for a normal operator $C$ in $L(\mathcal{H})$: there is a unitary operator $\mathcal{F} : \mathcal{H} \longrightarrow L^2(M,\mu)$, for some measure space $(M,\mu)$, such that $\mathcal{F} C \mathcal{F}^{-1}$ is multiplication by a complex measurable function on $M$.

Finally, let $A$ be an unbounded self-adjoint operator on $\mathcal{H}$. For $u$ in $\mathcal{D}(A)$,

$$\begin{aligned} \langle (i-A)u,(i-A)u\rangle &= \langle u,u\rangle + i\langle u,Au\rangle - i\langle Au,u\rangle + \langle Au,Au\rangle \\ &= \langle u,u\rangle + \langle Au,Au\rangle \geq \langle u,u\rangle . \end{aligned}$$

Therefore $i-A$ is injective and since $i-A$ is a closed operator, this also shows that $\mathcal{R}(i-A)$ is closed. We also claim that $\mathcal{R}(i-A)$ is dense. If not, there is a non-zero vector $z$ orthogonal to it, by the projection theorem. By the definition of adjoint, $z$ is in the domain of $A^* = A$ and

$$0 = \langle z, iu-Au\rangle = -\langle A^*z,u\rangle + i\langle z,u\rangle .$$

This holds for all $u$ in $\mathcal{D}(A)$, so we may set $u = z$ and obtain

$$0 = -\langle Az,z\rangle + i\langle z,z\rangle ,$$

which is impossible since $\langle Az,z\rangle$ is real. Therefore $\mathcal{R}(i-A)$, being closed and dense, is all of $\mathcal{H}$, and since $\|(i-A)u\| \geq \|u\|$ for $u$ in $\mathcal{D}(A)$, $(i-A)^{-1}$ is in $L(\mathcal{H})$ with norm $\leq 1$. We claim that it is normal. Now $(i-A)^{-1*} = (i-A)^{*-1}$ and since $i$ is in $L(\mathcal{H})$, $(i-A)^* = i^*-A^* = -i-A$, so that $(i-A)^{-1*} = (-i-A)^{-1}$. But $i-A$ and $-i-A$ commute, so

$$\begin{aligned}(i-A)^{-1}(i-A)^{-1*} &= (i-A)^{-1}(-i-A)^{-1} = ((-i-A)(i-A))^{-1} \\ &= ((i-A)(-i-A))^{-1} = (-i-A)^{-1}(i-A)^{-1} \\ &= (i-A)^{-1*}(i-A)^{-1}\end{aligned}$$

and $(i-A)^{-1}$ is normal. Hence there is a measure space $(M,\mu)$ and a unitary mapping $\mathcal{F} : \mathcal{H} \to L^2(M,\mu)$ taking $C = (i-A)^{-1}$ into multiplication by some complex measurable function $\hat{C}$. The function $\hat{C}$ is different from 0 almost everywhere because $C = (i-A)^{-1}$ does not annihilate any non-zero vector. Therefore $\hat{C}^{-1}$ is a complex measurable function, and so is $\hat{A} = i-\hat{C}^{-1}$. Thus $\mathcal{F}$ takes $A$ into multiplication by $\hat{A}$. Since multiplication by $\hat{A}$ is self-adjoint, $\hat{A}$ is real almost everywhere. This proves the spectral theorem for an unbounded self-adjoint operator.

To conclude the proof of the theorem, we need only show the converse. More generally, if $f$ is a complex measurable function on the measure space $(M,\mu)$ we will show that $M_f^* = M_{\bar{f}}$. Let $g$ be in $\mathcal{D}(M_f^*)$. Then there is an $h$ in $L^2(M,\mu)$ (in fact, $h = M_f^*g$) such that

$$\int \overline{h(x)}k(x)d\mu(x) = \int \overline{g(x)}f(x)k(x)d\mu(x)$$

for all $k$ in $\mathcal{D}(M_f)$. From this it follows that $h = \bar{f}g$ a.e., so that $M_f^* \subset M_{\bar{f}}$. The reverse inclusion is obvious. QED

As in the case of a self-adjoint operator in $L(\mathcal{H})$, if $A$ is an arbitrary self-adjoint operator and $f$ is a Baire function we may use the spectral theorem to define the operator $f(A)$, and the definition is independent of the choice of unitary map $\mathcal{F}$ and measure space $(M,\mu)$.

Theorem 4 (Stone's theorem). Let $U(t)$ be a strongly continuous one-parameter unitary group on a Hilbert space $\mathcal{H}$. Then there is a unique self-adjoint operator $A$ such that

$$U(t) = e^{itA} . \tag{4}$$

Conversely, if $A$ is a self-adjoint operator then (4) is a strongly continuous one-parameter unitary group.

Proof. The converse follows easily from the spectral theorem.

Let $U(t)$ be a strongly continuous one-parameter unitary group. Define its infinitesimal generator $B$ by

$$Bu = \lim_{h \to 0} \frac{U(h)-1}{h} u$$

on the domain $\mathcal{D}(B)$ of all $u$ for which the limit exists. We will show that $B$ is skew-adjoint; that is, that $B^* = -B$.

Let $\operatorname{Re} \lambda > 0$ and let

$$R_\lambda u = \int_0^\infty e^{-\lambda t} U(t) u\, dt , \qquad u \in \mathcal{H} .$$

Clearly, $R_\lambda$ is in $L(\mathcal{H})$ with norm $\leq 1/\operatorname{Re} \lambda$. We have

$$\frac{U(h)-1}{h} R_\lambda u = \frac{1}{h}\{\int_0^\infty e^{-\lambda t}U(t+h)udt - \int_0^\infty e^{-\lambda t}U(t)udt\}$$

$$= \int_0^\infty \frac{(e^{-\lambda(t-h)}-e^{-\lambda t})}{h} U(t)udt - \frac{1}{h}\int_0^h e^{-\lambda(t-h)}U(t)udt$$

$$\longrightarrow \lambda \int_0^\infty e^{-\lambda t}U(t)udt - u = \lambda R_\lambda u - u .$$

Therefore $\mathcal{R}(R_\lambda) \subset \mathcal{D}(B)$ and $(\lambda-B)R_\lambda = 1$ .

As $\lambda \longrightarrow \infty$ , $\lambda$ real, $\lambda R_\lambda$ converges strongly to 1 since $\lambda e^{-\lambda t}$ has integral 1 and becomes concentrated at 0 (and $\|U(t)u\|$ is bounded). Since each $\lambda R_\lambda u$ is in $\mathcal{D}(B)$ and $\lambda R_\lambda u \longrightarrow u$ , the operator $B$ is densely defined.

If $u$ is in $\mathcal{D}(B)$ it is clear that $U(t)u$ is in $\mathcal{D}(B)$ and $BU(t)u = U(t)Bu$ . Hence if $u$ is in $\mathcal{D}(B)$, $U(t)u$ is differentiable. We have

$$\frac{d}{dt} \langle U(t)u,U(t)u\rangle\Big|_{t=0} = \langle Bu,u\rangle + \langle u,Bu\rangle = 2\operatorname{Re}\langle u,Bu\rangle .$$

But $\langle U(t)u,U(t)u\rangle$ is a constant since $U(t)$ is unitary, so that $\operatorname{Re}\langle u,Bu\rangle = 0$ . Consequently $B$ is skew-Hermitean; that is, $B \subset -B^*$ .

Now let $u$ be in $\mathcal{D}(B)$ . Then

$$R_\lambda Bu = \int_0^\infty e^{-\lambda t}U(t)\,Bu\,dt = \int_0^\infty e^{-\lambda t}\frac{d}{dt} U(t)udt$$

$$= -\int_0^\infty (\frac{d}{dt} e^{-\lambda t})U(t)udt + e^{-\lambda t}U(t)u\Big|_0^\infty = \lambda R_\lambda u - u$$

so that $R_\lambda(\lambda-B)u = u$ for $u$ in $\mathcal{D}(B)$ and $\operatorname{Re}\lambda > 0$ . Together with $(\lambda-B)R_\lambda = 1$ , this implies that

$$R_\lambda = (\lambda-B)^{-1} , \qquad \operatorname{Re}\lambda > 0 .$$

If we replace $t$ by $-t$ we find in the same way that

$$\int_0^\infty e^{-\lambda t} U(-t) u dt = (\lambda + B)^{-1} u \ .$$

But the left hand side is $R^*_{\bar\lambda} u$ , so that

$$R^*_\lambda = (\bar\lambda + B)^{-1} \ .$$

Thus

$$(\lambda - B)^{-1*} = (\bar\lambda + B)^{-1} \ ,$$

so that

$$(\lambda - B)^* = \bar\lambda + B$$

and since $\lambda$ is in $L(\mathcal{H})$ , $-B^* = B$ and $B$ is skew-adjoint. Consequently $A = -iB$ is self-adjoint.

Let

$$V(t) = e^{itA} \ .$$

To conclude the proof we need only show that $V(t) = U(t)$ . One way to do this is to observe that for $\operatorname{Re} \lambda > 0$ ,

$$\int_0^\infty e^{-\lambda t} V(t) u dt = (\lambda - iA)^{-1} u = \int_0^\infty e^{-\lambda t} U(t) u dt \ .$$

By the uniqueness theorem for Laplace transforms, $V(t) = U(t)$ . A more direct proof is the following.

For $c < \infty$ let $\mathcal{H}_c$ be the set of all $u$ in $\mathcal{D}(A^n)$ for all $n$ such that

$$\|A^n u\| \le c^n \|u\|$$

for all $u$ . By the spectral theorem, $\mathcal{H}_c$ is a closed linear subspace of $\mathcal{H}$ invariant under $A$ , and the union of the $\mathcal{H}_c$ is dense in $\mathcal{H}$. (The space $\mathcal{H}_c$ is carried by $\mathcal{F}$ onto the space of all $L^2$ functions which vanish almost everywhere outside the set where $|\hat{A}| \leq c$ .) Since $A^n U(t)u = U(t)A^n u$ for $u$ in $\mathcal{H}_c$, $U(t)$ leaves $\mathcal{H}_c$ invariant. On $\mathcal{H}_c$ both $U(t)$ and $V(t)$ satisfy the differential equation that their derivative is $iA$, which is in $L(\mathcal{H}_c)$. Therefore by the uniqueness assertion of Theorem 4, §2, $U(t) = V(t)$ on $\mathcal{H}_c$. Since the union of the $\mathcal{H}_c$ is dense, $U(t) = V(t)$. QED

## 6. Commutative multiplicity theory

An unsatisfactory aspect of the spectral theorem as we have presented it is the lack of uniqueness in the choice of the measure space $(M,\mu)$ and the unitary transformation $\mathcal{F}$. In this section we will study the problem more thoroughly and obtain a complete classification of self-adjoint operators up to unitary equivalence. On a finite dimensional Hilbert space this is easy: two self-adjoint operators are unitarily equivalent if and only if they have the same eigenvalues with the same multiplicities.

Multiplicity theory for unbounded self-adjoint operators is essentially the same as for bounded self-adjoint operators, and without any genuine increase of difficulty of the problem we may study families of commuting self-adjoint operators.

In our treatment we shall use the notion of $\sigma$-function due to Paul Lévy and Laurent Schwartz [18]. This is a concept which is useful in other contexts and deserves to be more widely known.

A $C^*$-algebra is a norm-closed subalgebra $\mathcal{A}$ of $L(\mathcal{H})$ such that $A$ in $\mathcal{A}$ implies that $A^*$ is in $\mathcal{A}$. The following theorem is due to Stone.

Theorem 1. *Let $\mathcal{A}$ be a commutative $C^*$ algebra containing 1. Then there is a compact Hausdorff space $X$ such that $\mathcal{A}$ is * isomorphic and isometric to $C(X)$. The space $X$ is unique up to homeomorphism.*

If $\varphi: \mathcal{A} \to C(X)$ is the isomorphism, to say that it is a * isomorphism means that $\varphi(A^*) = \overline{\varphi(A)}$. The space $X$ is called the spectrum of $\mathcal{A}$ (or Gelfand maximal ideal space of $\mathcal{A}$).

Proof. As we saw in the course of the proof of the spectral theorem, if $A_1,\ldots,A_n$ are commuting self-adjoint operators in $L(\mathcal{H})$ and $p$ is a polynomial in $n$ variables then

$$\|p(A_1,\ldots,A_n)\| \leq \sup_{\lambda \in J} |p(\lambda)| \tag{1}$$

where

$$J = \prod_{i=1}^{n} [-\|A_i\|,\|A_i\|] .$$

Let $\mathcal{A}_0$ be the set of all self-adjoint operators in $\mathcal{A}$ and let

$$I = \prod_{A \in \mathcal{A}_0} [-\|A\|,\|A\|] .$$

Thus an element $x$ of $I$ is a function $x: A \rightsquigarrow x_A$ from $\mathcal{A}_0$ to

$$\bigcup_{A \in \mathcal{A}_0} [-\|A\|,\|A\|]$$

such that for each $A$ in $\mathcal{A}_0$, $x_A$ is in $[-\|A\|,\|A\|]$. With the product topology, $I$ is a compact Hausdorff space by the Tychonoff theorem [20]. For $A$ in $\mathfrak{a}_0$, let $\lambda_A$ be the function in $C(I)$ given by

$$\lambda_A(x) = x_A ,$$

and let $C_f(I)$ be the subalgebra of $C(I)$ generated by the functions $\lambda_A$. By the Stone-Weierstrass theorem [20, p.8], $C_f(I)$ is dense in $C(I)$. Any element of $C_f(I)$ is of the form $p(\lambda_{A_1},\ldots,\lambda_{A_n})$ for some $A_1,\ldots,A_n$ in $\mathfrak{a}_0$ and some polynomial $p$ in $n$ variables. By (1), the * homomorphism

$$p(\lambda_{A_1},\ldots,\lambda_{A_n}) \rightsquigarrow p(A_1,\ldots,A_n)$$

from $C_f(I)$ to $\mathfrak{a}$ is norm-decreasing, and so has a unique extension to a norm-decreasing * homomorphism

$$\varphi \colon C(I) \longrightarrow \mathfrak{a} .$$

Let $\mathfrak{n}$ be the kernel of $\varphi$ and let $X$ be the "hull" of $\mathfrak{n}$; that is,

$$X = \{x \in I \colon f(x) = 0 \text{ for all } f \text{ in } \mathfrak{n}\} .$$

Now $\mathfrak{n} \subset C(I-X)$; that is, the restrictions of functions in $\mathfrak{n}$ to the looally compact Hausdorff space $I-X$ are continuous and vanish at infinity. There is no point in $I-X$ at which all of the functions in $\mathfrak{n}$ vanish, and since $\mathfrak{n}$ is an ideal in $C(I)$ it separates points of $I-X$. By the Stone-Weierstrass theorem, $\mathfrak{n}$ is dense in $C(I-X)$. Since it is closed in $C(I)$ it is complete, and consequently $\mathfrak{n} = C(I-X)$. That is,

$$\mathfrak{N} = \{f \in C(I): f(x) = 0 \text{ for all } x \text{ in } X\} .$$

Restriction to $X$ gives a homomorphism $C(I) \longrightarrow C(X)$ whose kernel is $\mathfrak{N}$. By the Tietze extension theorem this mapping is surjective. Hence $C(I)/\mathfrak{N}$ is isomorphic to $C(X)$. But $C(I)/\mathfrak{N}$ is also isomorphic to $\mathfrak{A}$. Consequently there is an isomorphism $\varphi: C(X) \longrightarrow \mathfrak{A}$, which is clearly a * isomorphism.

We need to show that $\varphi$ is an isometry. If $f$ is in $C(X)$ then $\|f\|^2 - f\bar{f} \geq 0$, so for some $g$ in $C(X)$

$$\|f\|^2 - f\bar{f} = g\bar{g} .$$

Consequently

$$\|f\|^2 - \varphi(f)\varphi(f)^* = \varphi(g)\varphi(g)^* \geq 0 ,$$

so that $\|\varphi(f)\| \leq \|f\|$. We may work the same argument backwards. If $A$ is in $\mathfrak{A}$ then $\|A\|^2 = \|AA^*\|$, and $\|A\|^2 - AA^* \geq 0$, so for some $B$ in $\mathfrak{A}$

$$\|A\|^2 - AA^* = BB^* .$$

(We know that $B = (\|A\|^2 - AA^*)^{\frac{1}{2}}$ is in $\mathfrak{A}$ since the square root function is continuous on $[0,\|A\|^2]$ and therefore uniformly approximable by polynomials.) Consequently

$$\|A\|^2 - \varphi^{-1}(A)\,\overline{\varphi^{-1}(A)} = \varphi^{-1}(B)\overline{\varphi^{-1}(B)} \geq 0 ,$$

so that $\|\varphi^{-1}(A)\| \leq \|A\|$. Thus both $\varphi$ and $\varphi^{-1}$ are norm-decreasing, and $\varphi$ is an isometry.

By the Riesz representation theorem every continuous linear functional on $C(X)$ is given by a complex measure $\mu$ on $X$, and if the functional is multiplicative it is easy to see that $\mu$ must be the unit mass at some point $x$ in $X$. The topology of $X$ agrees with the weak-* topology of the corresponding multiplicative linear functionals. Thus the compact Hausdorff space $X$ is describable in terms of the Banach algebra $C(X)$, so that $X$ is unique up to homeomorphism. QED

A representation $\rho$ of $C(X)$, for $X$ a compact Hausdorff space, is a * homomorphism $\rho: C(X) \longrightarrow L(\mathcal{H})$ with $\rho(1) = 1$, where $\mathcal{H}$ is a Hilbert space. The space $\mathcal{H}$ is called the representation space of $\rho$ and is denoted by $\mathcal{H}(\rho)$. The argument used above to show that $\varphi$ was an isometry shows that any representation is norm-decreasing. Two representations $\rho_1$ and $\rho_2$ of $C(X)$ on Hilbert spaces $\mathcal{H}_1$ and $\mathcal{H}_2$ are unitarily equivalent, $\rho_1 \sim \rho_2$, in case there is a unitary operator $U: \mathcal{H}_1 \longrightarrow \mathcal{H}_2$ such that

$$\rho_2(f) = U\rho_1(f)U^{-1}$$

for all $f$ in $C(X)$. We will classify all representations of $C(X)$ up to unitary equivalence. In this way we will find all commutative $C^*$ algebras, up to unitary equivalence. Two $C^*$ algebras $\mathcal{A}_1$ and $\mathcal{A}_2$ on Hilbert spaces $\mathcal{H}_1$ and $\mathcal{H}_2$ are called unitarily equivalent in case there is a unitary operator $U: \mathcal{H}_1 \longrightarrow \mathcal{H}_2$ such that

$$\mathcal{A}_2 = U\, \mathcal{A}_1 U^{-1} .$$

Notice that two commutative $C^*$ algebras can be * isomorphic and isometric without being unitarily equivalent; for example, the algebras of all scalars on Hilbert spaces of different dimensions.

A representation $\rho$ of $C(X)$ is <u>cyclic</u> in case there is a $z$ in $\mathcal{H}(\rho)$ such that the set of all $\rho(f)z$ with $f$ in $C(X)$ is dense in $\mathcal{H}(\rho)$, in which case $z$ is called a <u>cyclic vector</u> for $\rho$.

If $\mu$ is a measure on $X$ then $f \rightsquigarrow M_f$, where $M_f$ is multiplication by $f$ on $L^2(\mu)$, is a representation of $C(X)$. We shall denote it by $\rho_\mu$.

<u>Theorem</u> 2. <u>If</u> $\mu$ <u>is a measure on</u> $X$ <u>then</u> $\rho_\mu$ <u>is a cyclic representation of</u> $C(X)$. <u>Conversely, every cyclic representation</u> $\rho$ <u>of</u> $C(X)$ <u>is unitarily equivalent to</u> $\rho_\mu$ <u>for some measure</u> $\mu$ <u>on</u> $X$.

<u>Proof.</u> The function 1 is a cyclic vector for $\rho_\mu$. To prove the converse, let $\rho$ be a cyclic representation of $C(X)$ with cyclic vector $z$. Then $\mu(f) = \langle z, \rho(f)z\rangle$ is a positive linear functional on $C(X)$ and so is a measure on $X$. For $f$ in $C(X)$ define $Uf = \rho(f)z$. We have

$$\|Uf\|^2 = \langle\rho(f)z, \rho(f)z\rangle = \langle z, \rho(\bar{f}f)z\rangle = \int |f|^2 d\mu .$$

In particular, if $f = 0$ a.e. $[\mu]$ then $Uf = 0$. Thus $U$ is an isometry from the dense linear subspace $C(X)$ of $L^2(\mu)$ onto a dense linear subspace of $\mathcal{H}(\rho)$, and so extends uniquely to a unitary operator $U\colon L^2(\mu) \longrightarrow \mathcal{H}(\rho)$. Clearly $U$ is a unitary equivalence between $\rho_\mu$ and $\rho$. QED

If $\rho_1$ and $\rho_2$ are representations of $C(X)$, we say that $\rho_1$ is a <u>subrepresentation</u> of $\rho_2$ in case $\mathcal{H}(\rho_1) \subset \mathcal{H}(\rho_2)$ and

$$\rho_1(f)u = \rho_2(f)u$$

for all $f$ in $C(X)$ and $u$ in $\mathcal{H}(\rho_1)$. We say that $\rho_1$ is contained in $\rho_2$, $\rho_1 \subset \rho_2$, in case $\rho_1$ is unitarily equivalent to a subrepresentation of $\rho_2$.

If $\mu_1$ and $\mu_2$ are measures on $X$ we say that $\mu_1$ is absolutely continuous with respect to $\mu_2$, $\mu_1 \ll \mu_2$, in case for all Borel sets $E$ in $X$, $\mu_2(E) = 0$ implies $\mu_1(E) = 0$. By the Radon-Nikodym theorem, this is equivalent to saying that there is a unique positive element $d\mu_1/d\mu_2$ of $L^1(\mu_2)$ such that

$$\int f\mu_1 = \int f \frac{d\mu_1}{d\mu_2}\, d\mu_2$$

for all $f$ in $C(X)$ (or for all positive Borel-measurable functions $f$). The measures $\mu_1$ and $\mu_2$ are called equivalent, $\mu_1 \approx \mu_2$, in case $\mu_1 \ll \mu_2$ and $\mu_2 \ll \mu_1$.

Theorem 3. Let $\mu$ and $\nu$ be measures on $X$. Then $\rho_\mu \subset \rho_\nu$ if and only if $\mu \ll \nu$, and $\rho_\mu \sim \rho_\nu$ if and only if $\mu \approx \nu$.

Proof. If $\mu \ll \nu$, let

$$Uf = f\sqrt{\frac{d\mu}{d\nu}}$$

for all $f$ in $L^2(\mu)$. Then $U$ is unitary from $L^2(\mu)$ to a subspace of $L^2(\nu)$, and for all $g$ in $C(X)$,

$$\rho_\nu(g)Uf = gf\sqrt{\frac{d\mu}{d\nu}} = U\rho_\mu(g)f\,, \qquad f \in L^2(\mu)\,.$$

Thus $\rho_\mu \subset \rho_\nu$. If $\mu \approx \nu$ then the range of $U$ is $L^2(\nu)$ and $\rho_\mu \sim \rho_\nu$.

Conversely, suppose that $\rho_\mu \subset \rho_\nu$ and let $U$ be a unitary equivalence of $\rho_\mu$ with a subrepresentation of $\rho_\nu$. Then $|U1|^2$ is a positive function in $L^1(\nu)$ and for all $h \geq 0$ in $C(X)$ (so that $h = |f|^2$ with $f$ in $C(X)$),

$$\int h d\mu = \int |f|^2 d\mu = \int |Uf|^2 d\nu = \int |Uf\cdot 1|^2 d\nu = \int |fU1|^2 d\nu = \int |f|^2 |U1|^2 d\nu .$$

Consequently, $\mu \ll \nu$ and $|U1|^2 = d\mu/d\nu$ . If $\rho_\mu \sim \rho_\nu$ then $|U1|^2 > 0$ a.e. $[\nu]$ and $\mu \approx \nu$ . QED

Next we shall construct the Hilbert space $\mathcal{H}(X)$ of σ-functions on $X$ . We write $(f,\mu)$ for a pair with $f$ in $L^2(\mu)$ and $\mu$ a measure on $X$ . We say that $(f,\mu)$ and $(g,\nu)$ are equivalent in case for some $\lambda$ with $\mu \ll \lambda$ and $\nu \ll \lambda$ ,

$$f\sqrt{\frac{d\mu}{d\lambda}} = g\sqrt{\frac{d\nu}{d\lambda}} \quad \text{a e. } [\lambda] .$$

If this holds for some $\lambda$ and we also have a measure $\lambda'$ with $\mu \ll \lambda'$ and $\nu \ll \lambda'$ then we claim that we also have

$$f\sqrt{\frac{d\mu}{d\lambda'}} = g\sqrt{\frac{d\nu}{d\lambda'}} \quad \text{a.e. } [\lambda'] .$$

To see this, let $\lambda''$ be a measure with $\lambda \ll \lambda''$ and $\lambda' \ll \lambda''$ (for example, we may take $\lambda'' = \lambda+\lambda'$) . Then

$$f\sqrt{\frac{d\mu}{d\lambda}} = g\sqrt{\frac{d\nu}{d\lambda}} \qquad \text{a.e. } [\lambda] ,$$

$$f\sqrt{\frac{d\mu}{d\lambda}}\sqrt{\frac{d\lambda}{d\lambda''}} = g\sqrt{\frac{d\nu}{d\lambda}}\sqrt{\frac{d\lambda}{d\lambda''}} \qquad \text{a.e. } [\lambda''] ,$$

$$f\sqrt{\frac{d\mu}{d\lambda''}} = g\sqrt{\frac{d\nu}{d\lambda''}} \qquad \text{a.e. } [\lambda''] ,$$

$$f\sqrt{\frac{d\mu}{d\lambda'}}\sqrt{\frac{d\lambda'}{d\lambda''}} = g\sqrt{\frac{d\nu}{d\lambda'}}\sqrt{\frac{d\lambda'}{d\lambda''}} \qquad \text{a.e. } [\lambda''] ,$$

$$f\sqrt{\frac{d\mu}{d\lambda'}} = g\sqrt{\frac{d\nu}{d\lambda'}} \qquad \text{a.e. } [\lambda'] ,$$

with each equation implying the succeeding equation. It follows from this that equivalence is an equivalence relation. A <u>$\sigma$-function</u> is an equivalence class of pairs $(f,\mu)$ , the equivalence class of a pair $(f,\mu)$ being denoted by $f\sqrt{d\mu}$ . The set of all $\sigma$-functions on $X$ is denoted by $\mathcal{H}(X)$ .

We add $\sigma$-functions by defining

$$f\sqrt{d\mu} + g\sqrt{d\nu} = \left( f\sqrt{\frac{d\mu}{d\lambda}} + g\sqrt{\frac{d\nu}{d\lambda}} \right) \sqrt{d\lambda}$$

where $\mu << \lambda$ and $\nu << \lambda$ . This is independent of the choice of representatives $(f,\mu)$ and $(g,\nu)$ and of the choice of $\lambda$ . We define scalar multiplication by $a(f\sqrt{d\mu}) = (af)\sqrt{d\mu}$ . Then $\mathcal{H}(X)$ forms a vector space. We define an inner product on $\mathcal{H}(X)$ by

$$<f\sqrt{d\mu},\ g\sqrt{d\nu}> = \int \bar{f}g\sqrt{\frac{d\mu}{d\lambda}}\ \sqrt{\frac{d\nu}{d\lambda}}\, d\lambda$$

where $\mu << \lambda$ and $\nu << \lambda$ . This is well-defined, is independent of the choice of $\lambda$ , and gives a strictly positive sesquilinear form on $\mathcal{H}(X)$ .

We claim that $\mathcal{H}(X)$ is a Hilbert space. It remains only to show completeness. Let $f_n\sqrt{d\mu_n}$ be a Cauchy sequence. There is a $\lambda$ such that $\mu_n << \lambda$ for each $\mu_n$ in the sequence: it suffices to let

$$\lambda = \sum_n \frac{1}{2^n} \frac{\mu_n}{\mu_n(X)} .$$

Then $f_n\sqrt{\frac{d\mu_n}{d\lambda}}$ is a Cauchy sequence in $L^2(\lambda)$ and so has a limit $f$ in $L^2(\lambda)$ . Then $f\sqrt{d\lambda}$ is the limit of $f_n\sqrt{d\mu_n}$ in $\mathcal{H}(X)$ , and $\mathcal{H}(X)$ is a Hilbert space.

Notice that if $\mu$ is any measure on $X$ then $f \rightsquigarrow f\sqrt{d\mu}$ gives an isometric embedding of $L^2(\mu)$ into $\mathcal{H}(X)$. If $\mu \ll \nu$ then the image of $L^2(\mu)$ is isometrically embedded in the image of $L^2(\nu)$. The Hilbert space $\mathcal{H}(X)$ canonically pieces together all of the $L^2$ spaces which one may form, avoiding some non-uniqueness which would otherwise be present in the discussion of commutative multiplicity theory. We shall denote the image of $L^2(\mu)$ in $\mathcal{H}(X)$ by $\mathcal{L}^2(\mu)$.

Two measures $\mu_1$ and $\mu_2$ on $X$ are called <u>singular</u> (with respect to each other) in case there is a partition of $X$ into two Borel sets $X_1$ and $X_2$ such that $\mu_2(X_1) = \mu_1(X_2) = 0$. This is the same as saying that if $\nu \ll \mu_1$ and $\nu \ll \mu_2$ then $\nu = 0$. We write $\mu_1 \perp \mu_2$ in case $\mu_1$ and $\mu_2$ are singular.

<u>Theorem</u> 4. <u>If</u> $\mu$ <u>and</u> $\nu$ <u>are measures on</u> $X$ <u>then</u>

$$\mu \ll \nu \quad \underline{\text{if and only if}} \quad \mathcal{L}^2(\mu) \subset \mathcal{L}^2(\nu),$$
$$\mu \approx \nu \quad \underline{\text{if and only if}} \quad \mathcal{L}^2(\mu) = \mathcal{L}^2(\nu),$$
$$\mu \perp \nu \quad \underline{\text{if and only if}} \quad \mathcal{L}^2(\mu) \perp \mathcal{L}^2(\nu).$$

<u>Proof.</u> The proofs of the first two statements follow the proof of Theorem 3. The third statement is an equally easy exercise. QED

For all $h$ in $C(X)$ and $f\sqrt{d\mu}$ in $\mathcal{H}(X)$ define

$$\pi(h)f\sqrt{d\mu} = hf\sqrt{d\mu} .$$

This is well-defined and $\pi$ is a representation of $C(X)$. If one defines the notion of a multiplicity-free representation, one can show that $\pi$ is the maximal multiplicity-free representation of $C(X)$. We will build all representations of $C(X)$, up to unitary equivalence, out of multiples of subrepresentations of $\pi$ .

Let $\mathcal{K}$ be the set of all closed linear subspaces of $\mathcal{H}(X)$ which are invariant under $\pi$ . Notice that if $\mathcal{M}$ is in $\mathcal{K}$ so is $\mathcal{M}^\perp$. Elements of $\mathcal{K}$ will be called, simply, invariant subspaces of $\mathcal{H}(X)$ .

If $\varphi = f\sqrt{d\mu}$ and $\psi = g\sqrt{d\nu}$ are in $\mathcal{H}(X)$ , we define their product $\varphi\psi$ to be the measure given by

$$\varphi\psi = fg\sqrt{\frac{d\mu}{d\lambda}}\sqrt{\frac{d\nu}{d\lambda}}\,d\lambda$$

where $\mu \ll \lambda$ and $\nu \ll \lambda$ . This is well-defined, and so is the complex conjugate $\bar{\varphi}$ given by

$$\bar{\varphi} = \bar{f}\sqrt{d\mu} \ .$$

We see that the scalar product $\langle\varphi,\psi\rangle$ is equal to $\int \bar{\varphi}\psi$ .

Theorem 5. Let $\varphi$ be in $\mathcal{H}(X)$ . The smallest invariant subspace of $\mathcal{H}(X)$ containing $\varphi$ is $\mathcal{L}^2(\bar{\varphi}\varphi)$ .

Proof. Let $\varphi = f\sqrt{d\mu}$ and let $d\kappa = \bar{\varphi}\varphi = |f|^2 d\mu$ . Let

$$g = \frac{f}{|f|}$$

with the understanding that $g = 0$ where $f = 0$ . Then $\varphi = g\sqrt{d\kappa}$ . Since $g \neq 0$ a.e. $[\kappa]$ -- in fact, $|g| = 1$ a.e. $[\kappa]$ -- $g$ is a cyclic vector for the representation of $C(X)$ on $L^2(\kappa)$ . Therefore the smallest invariant subspace of $\mathcal{H}(X)$ containing $\varphi$ contains $\mathcal{L}^2(\kappa) = \mathcal{L}^2(\bar{\varphi}\varphi)$ , and since this is clearly an invariant subspace of $\mathcal{H}(X)$ , the theorem is proved.

The following fact is basic in our development of commutative multiplicity theory.

<u>Theorem</u> 6. <u>If</u> $\mathcal{M}_1$ <u>and</u> $\mathcal{M}_2$ <u>are in</u> $\mathcal{K}$ <u>and</u> $\mathcal{M}_1 \cap \mathcal{M}_2 = 0$ <u>then</u> $\mathcal{M}_1 \perp \mathcal{M}_2$.

<u>Proof.</u> Suppose not. Then there are $\varphi_1$ in $\mathcal{M}_1$ and $\varphi_2$ in $\mathcal{M}_2$ with $\langle\varphi_1, \varphi_2\rangle \neq 0$. By Theorem 5, $\mathcal{L}^2(\bar{\varphi}_1\varphi_1) \subset \mathcal{M}_1$ and $\mathcal{L}^2(\bar{\varphi}_2\varphi_2) \subset \mathcal{M}_2$. By Theorem 4, $\bar{\varphi}_1\varphi_1$ and $\bar{\varphi}_2\varphi_2$ are not singular since $\mathcal{L}^2(\bar{\varphi}_1\varphi_1)$ and $\mathcal{L}^2(\bar{\varphi}_2\varphi_2)$ are not orthogonal. Therefore there is a non-zero measure $\mu$ with $\mu \ll \bar{\varphi}_1\varphi_1$ and $\mu \ll \bar{\varphi}_2\varphi_2$. By Theorem 4 again, $\mathcal{L}^2(\mu) \subset \mathcal{L}^2(\bar{\varphi}_1\varphi_1) \subset \mathcal{M}_1$ and $\mathcal{L}^2(\mu) \subset \mathcal{L}^2(\bar{\varphi}_2\varphi_2) \subset \mathcal{M}_2$, which is a contradiction. QED

Let $\mathcal{H}_\alpha$ be a Hilbert space for each $\alpha$ in an index set $J$. By the <u>direct sum</u> $\sum_\alpha \mathcal{H}_\alpha$ is meant the Hilbert space of all functions $\alpha \rightsquigarrow u_\alpha$ from $J$ to $\bigcup_\alpha \mathcal{H}_\alpha$ such that $u_\alpha$ is in $\mathcal{H}_\alpha$ for each $\alpha$ and $\|u\|^2 = \sum_\alpha \|u_\alpha\|^2 < \infty$. Notice that all but a countable number of components $u_\alpha$ must be $0$, for each $u$ in $\sum_\alpha \mathcal{H}_\alpha$. If all of the $\mathcal{H}_\alpha$ are equal to some fixed $\mathcal{H}$ and the cardinality of $J$ is $n$, the direct sum is denoted by $n\mathcal{H}$. If $\rho_\alpha$ is a representation of $C(X)$ on $\mathcal{H}_\alpha$, the <u>direct sum</u> of the representations $\rho_\alpha$ is the representation $\sum_\alpha \rho_\alpha$ on $\sum_\alpha \mathcal{H}_\alpha$ defined by $\rho(f) = \sum_\alpha \rho_\alpha(f)$. The operator sum converges strongly since $\|\rho_\alpha\| \leq 1$ for each $\alpha$. If all the $\rho_\alpha$ are equal to some fixed $\rho$ and the cardinality of $J$ is $n$, the direct sum is denoted by $n\rho$.

If $\rho_0$ is some representation of $C(X)$ and we let $\rho = \aleph_0\rho_0$ then $2\rho$ and $3\rho$ are unitarily equivalent even though $2 \neq 3$. However, we have the following result.

<u>Theorem</u> 7. <u>Let</u> $\mu$ <u>be a non-zero measure on</u> $X$ . <u>If</u> $n\rho_\mu$ <u>is unitarily equivalent to</u> $m\rho_\mu$ <u>then</u> $n = m$ .

<u>Proof.</u> Choose notation so that $n \leq m$ , and let $U\colon \mathcal{H}(n\rho_\mu) \longrightarrow \mathcal{H}(m\rho_\mu)$ be the unitary equivalence.

Suppose first that $n$ is infinite. Let $1_\alpha$ be the vector in $\mathcal{H}(n\rho_\mu) = n\mathcal{H}(\rho_\mu) = nL^2(\mu)$ with component $1$ in the $\alpha$'th place and all other components $0$ . The $n\rho_\mu(f)1_\alpha$ generate a dense subspace of $\mathcal{H}(n\rho_\mu)$ . Therefore the $U(n\rho_\mu(f)1_\alpha) = m\rho_\mu(f)U1_\alpha$ generate a dense subspace of $\mathcal{H}(m\rho_\mu)$ . Only countably many components of $U1_\alpha$ are non-zero, so $m \leq \aleph_0 n = n$ , and $n = m$ .

Now let $n$ be finite and let $U_{\beta\alpha}$ be the $\beta$'th component of $U1_\alpha$ . If $u$ in $\mathcal{H}(n\rho_\mu)$ has components $u_\alpha$ with $u_\alpha$ in $C(X)$ , then $Uu$ has components

$$\sum_\alpha U_{\beta\alpha}u_\alpha$$

since $U$ is a unitary equivalence. By continuity, this remains true for arbitrary components $u_\alpha$ in $L^2(\mu)$ . From the fact that $U$ is unitary we have

$$\int \sum_{\beta\alpha}\sum \overline{U}_{\beta\alpha}\overline{u}_\alpha \sum_\gamma U_{\beta\gamma}v_\gamma d\mu = \int \sum_\varepsilon \overline{u}_\varepsilon v_\varepsilon d\mu$$

for all $u_\alpha$ and $v_\gamma$ in $L^2(\mu)$ , and since they are arbitrary

$$\sum_\beta \overline{U}_{\beta\alpha}U_{\beta\gamma} = \delta_{\alpha\gamma} \quad \text{a.e. } [\mu] .$$

Therefore $U_{\beta\gamma}(x)$ is a unitary, and consequently square, matrix for almost every $x$ . Thus $n = m$ . QED

Suppose that for each cardinal $n$ we have a space $\mathcal{H}_n$ in $\mathcal{K}$, such that $\mathcal{H}_m \perp \mathcal{H}_n$ whenever $n \neq m$ and such that the $\mathcal{H}_n$ span $\mathcal{H}(X)$. Let $\pi | \mathcal{H}_n$ be the restriction of $\pi$ to $\mathcal{H}_n$. Any representation of the form $\rho = \Sigma\, n(\pi | \mathcal{H}_n)$ is called a <u>standard representation.</u> Our goal is to show that every representation is unitarily equivalent to a unique standard representation. We begin by showing uniqueness.

<u>Theorem</u> 8. <u>If</u> $\rho_1$ <u>and</u> $\rho_2$ <u>are unitarily equivalent standard representations of</u> $C(X)$, <u>then</u> $\rho_1 = \rho_2$.

<u>Proof.</u> Let $\rho_1 = \Sigma\, n(\pi | \mathcal{H}_n^1)$ and $\rho_2 = \Sigma\, n(\pi | \mathcal{H}_n^2)$. Suppose that $\mathcal{H}_n^1 \cap \mathcal{H}_m^2 \neq 0$. Then there is a non-zero measure $\mu$ with $\mathcal{L}^2(\mu) \subset \mathcal{H}_n^1 \cap \mathcal{H}_m^2$. Thus $n\mathcal{L}^2(\mu) \subset n\mathcal{H}_n^1$. Now $n\mathcal{L}^2(\mu)$ may be characterized as the set of all $\varphi$ in $\Sigma\, k\mathcal{H}_k^1$ such that $\bar{\varphi}\varphi << \mu$. Therefore, if $U$ is the unitary equivalence relating $\rho_1$ and $\rho_2$, $Un\mathcal{L}^2(\mu) = m\mathcal{L}^2(\mu)$ since the measure $\bar{\varphi}\varphi$ satisfies $\int f\bar{\varphi}\varphi = \langle \varphi, \rho_1(f)\varphi \rangle = \langle U\varphi, \rho_2(f)U\varphi \rangle$. By Theorem 7, $n = m$. Since the $\mathcal{H}_n^1$, $\mathcal{H}_m^2$ are each orthogonal and span $\mathcal{H}(X)$, it follows that $\mathcal{H}_n^1 = \mathcal{H}_n^2$. QED

Let $\rho$ be a representation of $C(X)$ on a Hilbert space $\hat{\mathcal{H}}$, and let $\hat{\mathcal{K}}$ be the set of closed invariant linear subspaces of $\hat{\mathcal{H}}$. A <u>foundation</u> for $\rho$ is a pair $(\mathcal{M}, \hat{\mathcal{M}})$ with $\mathcal{M}$ in $\mathcal{K}$ and $\hat{\mathcal{M}}$ in $\hat{\mathcal{K}}$ such that $(\pi | \mathcal{M}) \sim (\rho | \hat{\mathcal{M}})$ and such that if $\mathcal{N}$ is in $\mathcal{K}$ and $(\pi | \mathcal{N}) \subset (\rho | \hat{\mathcal{M}}^{\perp})$ then $\mathcal{N} \subset \mathcal{M}$.

The last condition is a maximality condition. In the case of a standard representation $\Sigma\, n(\pi | \mathcal{H}_n)$, one foundation is the pair

$(\mathfrak{M}, \hat{\mathfrak{M}})$ where $\mathfrak{M}$ is the span of the $\mathcal{H}_n$ with $n \geq 1$ and $\hat{\mathfrak{M}}$ is the set of vectors all of whose components, except the first, vanish (in each summand).

Theorem 9. If $\rho$ is a representation of $C(X)$ then $\rho$ has a foundation. If $\rho$ is not the $0$ representation then the foundation is not $(0,0)$.

Proof. Let $\rho$ be a representation of $C(X)$ on $\hat{\mathcal{H}}$. Let the $(\mathfrak{M}_\alpha, \hat{\mathfrak{M}}_\alpha)$ be a maximal family of pairs with each $\mathfrak{M}_\alpha \neq 0$, $\mathfrak{M}_\alpha \in \mathcal{K}$, $\hat{\mathfrak{M}}_\alpha \in \hat{\mathcal{K}}$, $\mathfrak{M}_\alpha \perp \mathfrak{M}_\beta$ and $\hat{\mathfrak{M}}_\alpha \perp \hat{\mathfrak{M}}_\beta$ whenever $\alpha \neq \beta$, and with $(\pi|\mathfrak{M}_\alpha) \sim (\rho|\hat{\mathfrak{M}}_\alpha)$. This exists by Zorn's lemma. Let $\mathfrak{M}$ be the span of the $\mathfrak{M}_\alpha$ and $\hat{\mathfrak{M}}$ the span of the $\hat{\mathfrak{M}}_\alpha$. We claim that $(\mathfrak{M}, \hat{\mathfrak{M}})$ is a foundation for $\rho$.

Clearly, $(\pi|\mathfrak{M}) \sim (\rho|\hat{\mathfrak{M}})$. Suppose that $\mathfrak{N}$ in $\mathcal{K}$ is such that $(\pi|\mathfrak{N}) \sim (\rho|\hat{\mathfrak{N}})$ for some $\hat{\mathfrak{N}}$ in $\hat{\mathcal{K}}$ with $\hat{\mathfrak{N}} \subset \hat{\mathfrak{M}}^\perp$. Let $\mathfrak{Z} = \mathfrak{N} \cap \mathfrak{M}^\perp$. Then $\mathfrak{Z}$ is in $\mathcal{K}$. Furthermore, $(\pi|\mathfrak{Z}) \sim (\rho|\hat{\mathfrak{Z}})$ for some $\hat{\mathfrak{Z}}$ in $\hat{\mathcal{K}}$ with $\hat{\mathfrak{Z}} \subset \hat{\mathfrak{N}} \subset \hat{\mathfrak{M}}^\perp$, which contradicts the maximality of our family unless $\mathfrak{Z} = 0$. Since $\mathfrak{N} \cap \mathfrak{M}^\perp = 0$, $\mathfrak{N}$ is orthogonal to $\mathfrak{M}^\perp$ by Theorem 6. Thus $\mathfrak{N} \subset \mathfrak{M}$ and $(\mathfrak{M}, \hat{\mathfrak{M}})$ is a foundation.

Suppose that $\rho$ is not the $0$ representation; that is, suppose that $\hat{\mathcal{H}} \neq 0$. Then there is a $u \neq 0$ in $\mathcal{H}$. Let $\mu$ be the measure on $X$ such that $\int f d\mu = \langle u, \rho(f)u \rangle$ for $f$ in $C(X)$. Then $\mu \neq 0$, and $\pi$ restricted to $\mathcal{L}^2(\mu)$ is unitarily equivalent to $\rho$ restricted to the cyclic subspace generated by $u$. Thus $(0,0)$ is not a foundation for $\rho$. QED

Theorem 10. Every representation of C(X) is unitarily equivalent to a unique standard representation.

Proof. The uniqueness was proved in Theorem 8.

Consider a representation $\rho$ of C(X) on $\hat{\mathcal{H}}$, and let $(\mathfrak{M}_0, \hat{\mathfrak{M}}_0)$ be a foundation for $\rho$. If $(\mathfrak{M}_\beta, \hat{\mathfrak{M}}_\beta)$ has been defined for all ordinals $\beta < \alpha$, let $(\mathfrak{M}_\alpha, \hat{\mathfrak{M}}_\alpha)$ be a foundation for $\rho$ restricted to

$$\Big(\sum_{\beta < \alpha} \hat{\mathfrak{M}}_\beta\Big)^\perp .$$

These foundations exist by Theorem 9. Also by Theorem 9, the span of the $\hat{\mathfrak{M}}_\beta$ with $\beta < \alpha$ is eventually the entire space $\hat{\mathcal{H}}$. Consequently, $(\mathfrak{M}_\alpha, \hat{\mathfrak{M}}_\alpha) = (0,0)$ for large enough ordinals $\alpha$.

It follows easily from the definition of foundation that $\mathfrak{M}_\beta \subset \mathfrak{M}_\alpha$ whenever $\alpha < \beta$.

Let

$$\mathcal{H}_0 = \mathfrak{M}_0^\perp$$

and

$$\mathcal{H}_\beta = \bigcap_{\gamma < \beta} \mathfrak{M}_\gamma \cap \mathfrak{M}_\beta^\perp$$

for $\beta > 0$. Clearly $\mathcal{H}_\beta$ is in $\mathcal{K}$. Suppose $\alpha < \beta$. Then

$$\mathcal{H}_\alpha \cap \mathcal{H}_\beta = \bigcap_{\gamma < \beta} \mathfrak{M}_\gamma \cap \mathfrak{M}_\alpha^\perp \subset \mathfrak{M}_\alpha \cap \mathfrak{M}_\alpha^\perp = 0 .$$

By Theorem 6, the $\mathcal{H}_\alpha$ are orthogonal. We claim that they span $\mathcal{H}(X)$. Suppose not. Then there is a $\varphi \neq 0$ such that $\varphi$ is orthogonal to all $\mathcal{H}_\alpha$. In particular, $\varphi \perp \mathcal{H}_0$ so that $\varphi$ is in $\mathfrak{M}_0$. It cannot happen that $\varphi$ is in all $\mathfrak{M}_\alpha$ since they are eventually $0$.

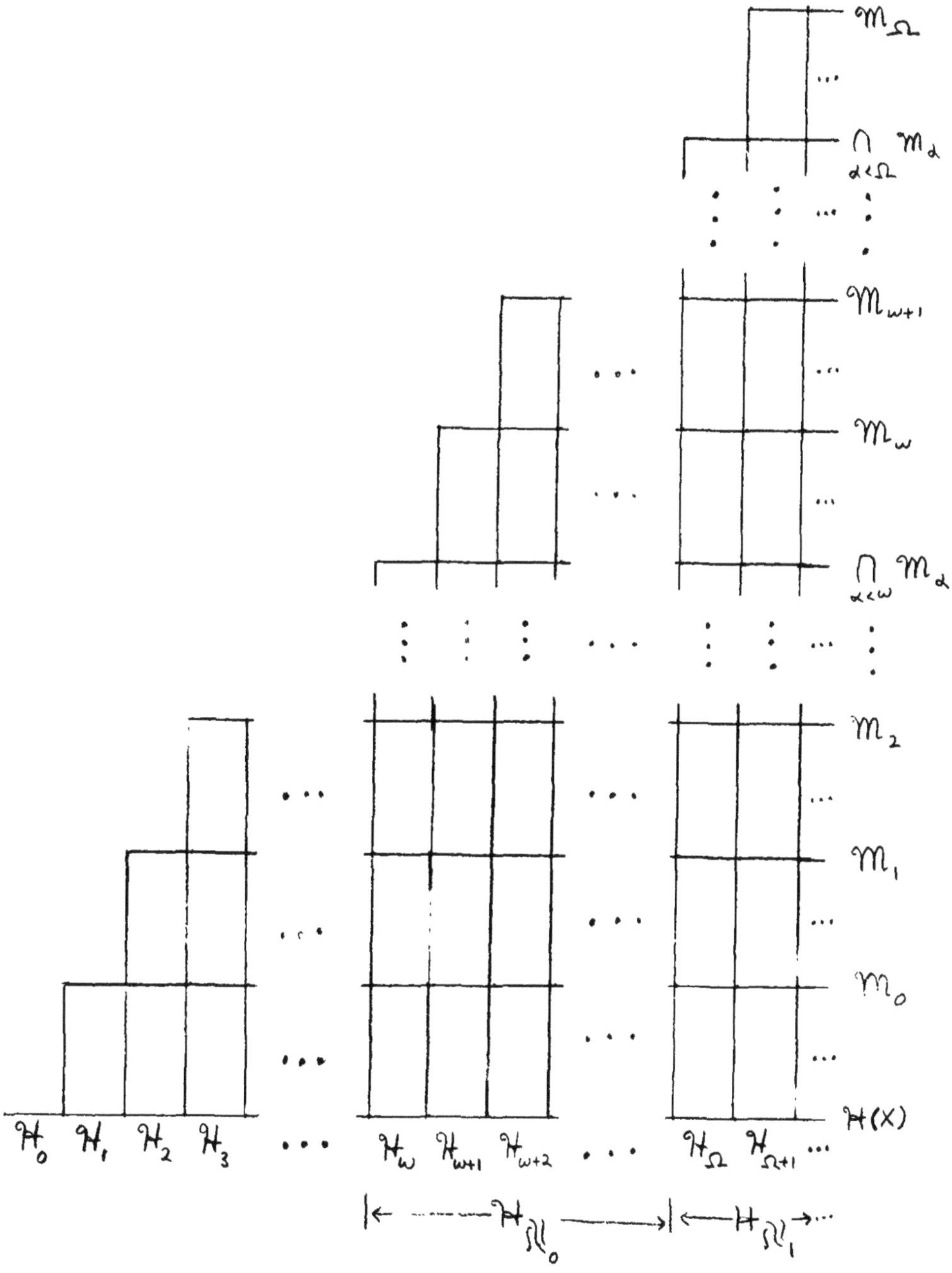


Figure 7.

Construction of the standard representation (Theorem 10).

Let $\delta$ be the least ordinal such that $\varphi$ is not in $\mathcal{M}_\delta$. Then $\varphi$ is in

$$\mathcal{N} = \bigcap_{\alpha < \delta} \mathcal{M}_\alpha ,$$

and $\varphi$ is orthogonal to $\mathcal{N} \cap \mathcal{M}_\delta^\perp$, since this is just $\mathcal{H}_\delta$. Therefore $\varphi$ is in $\mathcal{N} \cap \mathcal{M}_\delta$, which is a contradiction.

Now we shall construct the desired standard representation. If $\beta$ is an ordinal let $|\beta|$ be its cardinal. Let

$$\mathcal{H}_n = \sum_{|\beta|=n} \mathcal{H}_\beta .$$

Then the $\mathcal{H}_n$ are orthogonal and span $\mathcal{H}(X)$, so that $\Sigma\, n(\pi|\mathcal{H}_n)$ is a standard representation.

From the definition of the $\mathcal{H}_\beta$, we see that

$$\mathcal{M}_\alpha \cap \mathcal{H}_\beta = \mathcal{H}_\beta , \qquad \alpha < \beta \tag{2}$$

and that

$$\mathcal{M}_\alpha \cap \mathcal{H}_\beta = 0 , \qquad \alpha \geq \beta ,$$

and consequently by Theorem 6,

$$\mathcal{M}_\alpha \perp \mathcal{H}_\beta , \qquad \alpha \geq \beta . \tag{3}$$

Now $\rho$ is clearly unitary equivalent to the direct sum of the restrictions of $\pi$ to $\mathcal{M}_\alpha$,

$$\rho \sim \sum_\alpha (\pi|\mathcal{M}_\alpha) .$$

By (2) and (3), and the fact that the $\mathcal{H}_\beta$ are orthogonal and span $\mathcal{H}(X)$,

$$\mathcal{M}_\alpha = \sum_\beta \mathcal{M}_\alpha \cap \mathcal{H}_\beta .$$

Therefore

$$\rho \sim \sum_{\alpha} \sum_{\beta} (\pi | \mathcal{M}_{\alpha} \cap \mathcal{H}_{\beta}) = \sum_{\beta} \sum_{\alpha} (\pi | \mathcal{M}_{\alpha} \cap \mathcal{H}_{\beta})$$

$$= \sum_{\beta} \sum_{\alpha < \beta} (\pi | \mathcal{H}_{\beta}) = \sum_{\beta} |\beta| (\pi | \mathcal{H}_{\beta}) = \sum_{n} n(\pi | \mathcal{H}_{n}) .$$

This completes the proof.

We may state some corollaries of the main theorem in a language not involving $\sigma$-functions. Let $\rho$ be a representation of $C(X)$. The <u>multiplicity function</u> of $\rho$ is the function $\mu \rightsquigarrow \text{mult}(\mu)$ from measures on $X$ to cardinals such that $\text{mult}(\mu)$ is the maximal cardinal number $m$ such that $m\rho_{\mu} \subset \rho$. For a standard representation $\sum_{n} (\pi | \mathcal{H}_{n})$ it is easy to see that $\text{mult}(\mu)$ is the maximal cardinal $m$ such that

$$\mathcal{L}^{2}(\mu) \subset \sum_{k \geq m} \mathcal{H}_{k} .$$

The following result is an easy corollary of this and Theorem 10.

<u>Theorem</u> 11. <u>Two representations of</u> $C(X)$ <u>are unitarily equivalent if and only if they have the same multiplicity function.</u>

Inseparable Hilbert spaces are of little interest. Suppose that $\rho$ is a representation of $C(X)$ on a separable Hilbert space $\hat{\mathcal{H}}$. Then the $\mathcal{H}_{n}$ in the corresponding standard representation are $0$ for $n > \aleph_{0}$. For $n \leq \aleph_{0}$, let the $\mu_{\alpha}$ be a maximal family of pairwise singular non-zero measures with $\mathcal{L}^{2}(\mu_{\alpha}) \subset \mathcal{H}_{n}$. Then $\mathcal{H}_{n}$ is the span of the orthogonal spaces $\mathcal{L}^{2}(\mu_{\alpha})$. There are only countably many of these $\mu_{\alpha}$, say $\mu_{1}, \mu_{2}, \ldots$. Let

$$\mu_{(n)} = \Sigma \frac{1}{2^k} \frac{\mu_k}{\mu_k(X)} .$$

This is a measure on $X$ and $\mathcal{L}^2(\mu_{(n)}) = \mathcal{H}_n$. The $\mu_{(n)}$ are pairwise singular. Let $\mu$ be the measure

$$\mu = \sum_{1 \leq n < \aleph_0} \frac{1}{2^n} \mu_{(n)} + \mu_{(\aleph_0)} .$$

Then we have the following theorem.

<u>Theorem</u> 12. <u>Let</u> $\rho$ <u>be a representation of</u> $C(X)$ <u>on a separable Hilbert space. Then there are disjoint Borel sets</u> $E_n$ <u>in</u> $X$, <u>for</u> $1 \leq n \leq \aleph_0$, <u>and a measure</u> $\mu$ <u>on</u> $X$ <u>such that</u> $\rho$ <u>is unitarily equivalent to the direct sum of</u> $n$ <u>times the multiplication representation of</u> $C(X)$ <u>on</u> $L^2(E_n, \mu)$. <u>Another such representation, with sets</u> $E'_n$ <u>and measure</u> $\mu'$, <u>is unitarily equivalent to the first if and only if the measures</u> $\mu$ <u>and</u> $\mu'$ <u>are equivalent and</u> $E_n = E'_n$ <u>a.e.</u>

Finally, we may classify self-adjoint operators. If $A$ is a self-adjoint operator on a Hilbert space, then arctan $A$ is a bounded self-adjoint operator. In fact, it has norm $\leq \frac{\pi}{2}$. The operator $A$ may be recovered from arctan $A$ since $A = \tan \arctan A$. (The only property of the arctan function we are using is that it is injective with bounded range.) To classify arctan $A$ we need only apply our above results concerning representations of $C(X)$ for $X = [-\frac{\pi}{2}, \frac{\pi}{2}]$.

<u>Theorem</u> 13. <u>Let</u> $A$ <u>be a self-adjoint operator on a separable Hilbert space</u> $\mathcal{H}$. <u>There are Borel sets</u> $E_n$ <u>in</u> $\mathbb{R}$, <u>for</u> $1 \leq n \leq \aleph_0$, <u>and a measure</u> $\mu$ <u>on</u> $\mathbb{R}$ <u>such that</u> $A$ <u>is unitarily</u>

equivalent to the direct sum of $n$ times the multiplication operator by the identity function on $L^2(E_n,\mu)$. Two self-adjoint operators $A$ and $A'$ on a separable Hilbert space are unitarily equivalent if and only if the corresponding measures $\mu$ and $\mu'$ are equivalent and the corresponding sets $E_n$ and $E'_n$ are equal a.e.

The notions of $\mathcal{H}(X)$ and multiplication by a continuous function on it may be defined in the obvious way for a locally compact Hausdorff space. We then have the following result.

Theorem 14. Let $A$ be a self-adjoint operator on a Hilbert space. For each cardinal $n$ there is a unique closed invariant subspace $\mathcal{H}_n$ of $\mathcal{H}(\mathbb{R})$ such that the $\mathcal{H}_n$ are orthogonal and span $\mathcal{H}(\mathbb{R})$ and such that $A$ is unitarily equivalent to the direct sum of $n$ times the multiplication operator by the identity function on $\mathcal{H}_n$.

## 7. Extensions of Hermitean operators

A Hermitean operator $A$ on a Hilbert space $\mathcal{H}$ is called essentially self-adjoint in case $\bar{A}$ is self-adjoint. A complex number $\lambda$ is in the resolvent set of an operator $A$ in case $\lambda - A$ is injective and $(\lambda - A)^{-1}$ is in $L(\mathcal{H})$.

Theorem 1. Let $A$ be a Hermitean operator on a Hilbert space. Then the following are equivalent:

(i) $A$ is essentially self-adjoint, $\bar{A}^* = \bar{A}$,

(ii) $\bar{A} = A^*$,

(iii) $A^* = A^{**}$,

(iv) $A^* \subset A^{**}$,

(v) $\mathcal{R}(i-A)$ <u>and</u> $\mathcal{R}(-i-A)$ <u>are dense,</u>

(vi) $i$ <u>and</u> $-i$ <u>are not eigenvalues of</u> $A^*$,

(vii) $i$ <u>and</u> $-i$ <u>are in the resolvent set of</u> $\bar{A}$,

(viii) $(i-\bar{A})^{-1}$ <u>is in</u> $L(\mathcal{H})$ <u>and is normal,</u>

(ix) $(-i-\bar{A})^{-1}$ <u>is in</u> $L(\mathcal{H})$ <u>and is normal.</u>

<u>Proof.</u> Since $\bar{A}^* = A^*$ and $\bar{A} = A^{**}$, (i), (ii), and (iii) are clearly equivalent. They imply (iv), but since $A$ is Hermitean, if (iv) holds then $A \subset \bar{A} \subset A^* \subset A^{**} = \bar{A}$, so that (iv) implies (iii). Thus (i) through (iv) are equivalent. By the spectral theorem, they imply (v) through (ix). If we use the fact that $\|(\pm i-A)u\| \geq \|u\|$ for $u$ in $\mathcal{D}(A)$ we see that (v) through (vii) are equivalent. By the argument given at the end of the proof of the spectral theorem (Theorem 3, §5), they imply (viii) and (ix), which imply (ii). QED

If $A$ is a Hermitean operator, the ordered pair of cardinal numbers

$$(\dim \mathcal{R}(-i-A)^\perp, \dim \mathcal{R}(i-A)^\perp)$$

is called the <u>deficiency indices</u> of $A$. Thus $A$ is essentially self-adjoint if and only if it has deficiency indices $(0,0)$.

If $A$ is in $L(\mathcal{H})$ then the sesquilinear form $(u,v) \rightsquigarrow \langle u,Av\rangle$ is bounded; that is, there is a constant $c < \infty$ such that

$$|\langle u,Av\rangle| \leq c\|u\|\|v\| ; \qquad u,v \in \mathcal{H} .$$

It is easy to see that every bounded sesquilinear form arises in this way. For unbounded operators and unbounded sesquilinear forms the relationship is more subtle and more interesting.

Let $\mathcal{H}$ be a Hilbert space and let $\mathcal{H}^1$ be a dense linear subspace which is itself a Hilbert space with a larger norm,

$$\|u\|_1 \geq \|u\| , \qquad u \in \mathcal{H}^1 .$$

Let $\mathcal{H}^{-1}$ be the set of all continuous linear functionals $u \rightsquigarrow \langle u, v\rangle$ on $\mathcal{H}^1$. (Notice that we use the same notation for the pairing between $\mathcal{H}^{-1}$ and $\mathcal{H}^1$ as for the inner product in $\mathcal{H}$.) We make $\mathcal{H}^{-1}$ into a vector space by defining addition by

$$\langle u_1 + u_2, v\rangle = \langle u_1, v\rangle + \langle u_2, v\rangle$$

and scalar multiplication by

$$\langle au, v\rangle = \bar{a}\langle u, v\rangle ,$$

and we give $\mathcal{H}^{-1}$ the norm

$$\|u\|_{-1} = \sup_{\|v\|_1 \leq 1} |\langle u, v\rangle| .$$

Then $\mathcal{H}^{-1}$ is a Banach space. By the Riesz representation theorem there is a unique bijective isometry

$$J: \mathcal{H}^1 \longrightarrow \mathcal{H}^{-1}$$

such that

$$\langle u, v\rangle_1 = \langle Ju, v\rangle , \qquad u, v \in \mathcal{H}^1 ,$$

where $\langle\ ,\ \rangle_1$ denotes the inner product in $\mathcal{H}^1$. The space $\mathcal{H}^{-1}$ is a Hilbert space with the inner product

$$\langle u,v\rangle_{-1} = \langle J^{-1}u, J^{-1}v\rangle_1 , \qquad u,v \in \mathcal{H}^{-1} .$$

Thus $J\colon \mathcal{H}^1 \longrightarrow \mathcal{H}^{-1}$ is unitary.

If $w$ is in $\mathcal{H}$ then $u \rightsquigarrow \langle w,u\rangle$, for $u$ in $\mathcal{H}^1$, is a linear functional, and since

$$|\langle w,u\rangle| \le \|w\|\|u\| \le \|w\|\|u\|_1$$

it is in $\mathcal{H}^{-1}$, with norm $\le \|w\|$. If it is the zero linear functional then $w = 0$, since $\mathcal{H}^1$ is dense in $\mathcal{H}$. Thus we have a natural injection of $\mathcal{H}$ into $\mathcal{H}^{-1}$ which diminishes norms. We shall simply identify $\mathcal{H}$ as a subspace of $\mathcal{H}^{-1}$. Thus we have

$$\mathcal{H}^{-1} \supset \mathcal{H} \supset \mathcal{H}^1 .$$

The space $\mathcal{H}^1$ (and consequently $\mathcal{H}$) is dense in $\mathcal{H}^{-1}$, for if not there is a $z$ in $\mathcal{H}^1$ which is orthogonal to all $u$ in $\mathcal{H}^1$:

$$\langle u,z\rangle = 0 , \qquad u \in \mathcal{H}^1 .$$

Since $\mathcal{H}^1$ is dense in $\mathcal{H}$, this means that $\langle u,z\rangle = 0$ for all $u$ in $\mathcal{H}$ and consequently $z = 0$.

<u>Theorem</u> 2. <u>Let</u> $\mathcal{H}$ <u>be a Hilbert space and let</u> $\mathcal{H}^1$ <u>be a dense linear subspace which is a Hilbert space in a larger norm. Let</u> $\mathcal{H}^{-1}$ <u>and</u> $J\colon \mathcal{H}^1 \longrightarrow \mathcal{H}^{-1}$ <u>be as above. Let</u> $J_0$ <u>be the restriction of</u> $J$ <u>to all</u> $u$ <u>in</u> $\mathcal{H}^1$ <u>such that</u> $Ju$ <u>is in</u> $\mathcal{H}$. <u>Then</u> $J_0$ <u>is a self-adjoint operator on</u> $\mathcal{H}$.

<u>Proof</u>. Since $J\colon \mathcal{H}^1 \longrightarrow \mathcal{H}^{-1}$ is bijective, $J_0\colon \mathcal{D}(J_0) \longrightarrow \mathcal{H}$ is bijective. Since

$$\|J_0 u\| \geq \|Ju\|_{-1} = \|u\|_1 \geq \|u\| ,$$

$J_0^{-1}$ is in $L(\mathcal{H})$ with $\|J_0^{-1}\| \leq 1$ . Since

$$\langle J_0^{-1}u, u\rangle = \langle J^{-1}u, JJ^{-1}u\rangle = \|J^{-1}u\|_1^2$$

is real, $J_0^{-1}$ is self-adjoint. Therefore $J_0$ is self-adjoint. QED

Suppose that we have a dense linear subspace $\mathcal{D}$ of a Hilbert space $\mathcal{H}$ and that on $\mathcal{D}$ we have a sesquilinear form $(u,v) \rightsquigarrow \langle u,v\rangle_1$ such that

$$\langle u,u\rangle_1 \geq \langle u,u\rangle , \qquad u \in \mathcal{D} .$$

Then we may define $\mathcal{H}^1$ to be the completion of $\mathcal{D}$ in this inner product. Any sequence in $\mathcal{D}$ which is a Cauchy sequence in the norm $\|\ \|_1$ is also a Cauchy sequence in the norm $\|\ \|$ , so that we have a natural norm-decreasing linear mapping $\mathcal{H}^1 \to \mathcal{H}$ which is the identity on $\mathcal{D}$ . However, this mapping need not be injective. An example is the Hilbert space $\mathcal{H} = L^2(\mathbb{R})$ , $\mathcal{D}$ all continuous functions with compact support, $\langle u,v\rangle_1 = \langle u,v\rangle + \overline{u(0)}v(0)$ .

Theorem 3 (Friedrichs extension theorem). Let A be a densely defined linear operator on the Hilbert space $\mathcal{H}$ such that

$$\langle u,Au\rangle \geq \langle u,u\rangle , \qquad u \in \mathcal{D}(A) .$$

Let $\mathcal{H}^1$ be the completion of $\mathcal{D}(A)$ in the inner product $\langle u,v\rangle_1 = \langle u,Av\rangle$ . Then the identity mapping $\mathcal{D}(A) \to \mathcal{H}$ extends by continuity to an injective norm-decreasing linear map $\mathcal{H}^1 \to \mathcal{H}$ , so that we may identify $\mathcal{H}^1$ as a dense linear subspace of $\mathcal{H}$ which is a Hilbert space in a larger norm. The operator $J_0$ of Theorem 2 is a self-adjoint extension of $A$.

Proof. Since the identity mapping $\mathcal{D}(A) \longrightarrow \mathcal{H}$ is norm-decreasing, it extends by continuity to a unique norm-decreasing mapping $\iota$ in $L(\mathcal{H}^1, \mathcal{H})$. For all $u$ and $v$ in $\mathcal{D}(A)$,

$$\langle u, v\rangle_1 = \langle u, Av\rangle = \langle \iota u, Av\rangle .$$

By continuity, this holds for all $u$ in $\mathcal{H}^1$ and $v$ in $\mathcal{D}(A)$. Therefore if $\iota u = 0$, $u$ is orthogonal in $\mathcal{H}^1$ to $\mathcal{D}(A)$ and so is $0$. Thus $\iota$ is injective. It is clear that $A \subset J_0$. QED

The operator $J_0$ is called the Friedrichs extension of $A$. The operator $A$ may have other self-adjoint extensions, but the Friedrichs extension is constructed in a canonical way and is of great importance in many applications. A Hermitean operator $A$ is called semi-bounded in case for some $c < \infty$,

$$\langle u, Au\rangle \geq -c\langle u, u\rangle , \qquad u \in \mathcal{D}(A) .$$

If $A$ is semi-bounded then $A+c+1$ satisfies the hypotheses of Theorem 3. If $J_0$ is the Friedrichs extension of $A+c+1$ then $J_0-c-1$ is a self-adjoint extension of $A$, called its Friedrichs extension. Thus every semi-bounded Hermitean operator has a natural self-adjoint extension.

## 8. Sums and Lie products of self-adjoint operators

A theorem of Paul Chernoff [23] gives us the result needed in order to discuss the one-parameter unitary group generated by the sum or Lie product of two self-adjoint operators. The natural context for the discussion is given by the notion of a contraction semigroup on a Banach space.

Let $\mathfrak{X}$ be a Banach space. A contraction semigroup on $\mathfrak{X}$ is a family of operators $P^t$ in $L(\mathfrak{X})$, for $0 \leq t < \infty$, such that $\|P^t\| \leq 1$, $P^0 = 1$, $P^tP^s = P^{t+s}$, and

(1) $$\lim_{t \to 0} P^t u = u , \qquad u \in \mathfrak{X} .$$

This is usually called a "contraction semigroup of class $(C_0)$", the last phrase referring to the strong continuity condition (1). However, we will deal only with such semigroups.

An example of a contraction semigroup is $e^{itA}$, considered for $t \geq 0$, where $A$ is a self-adjoint operator on a Hilbert space.

The infinitesimal generator of a contraction semigroup $P^t$ is the operator $A$ defined by

$$Au = \lim_{h \to 0} \frac{P^h - 1}{h} u ,$$

on the domain $\mathcal{D}(A)$ of all $u$ in $\mathfrak{X}$ for which the limit exists.

THEOREM 1. *Let $P^t$ be a contraction semigroup with infinitesimal generator $A$. Then $A$ is a closed, densely defined operator and for all $\lambda$ with $\operatorname{Re} \lambda > 0$, $\lambda$ is in the resolvent set of $A$, $\|(\lambda - A)^{-1}\| \leq 1/\operatorname{Re} \lambda$, and*

(2)
$$(\lambda - A)^{-1}u = \int_0^\infty e^{-\lambda t}P^t u\,dt$$

for all $u$ in $\mathfrak{X}$ .

Proof (cf. the proof of Stone's theorem, Theorem 4, §5). Let $\operatorname{Re}\lambda > 0$ . Then the integral in (2) clearly converges and defines an operator $R_\lambda$ in $L(\mathfrak{X})$ with $\|R_\lambda\| \le 1/\operatorname{Re}\lambda$ . We have

$$\frac{P^h - 1}{h}u = \{\int_0^\infty e^{-\lambda t}P^{t+h}u\,dt - \int_0^\infty e^{-\lambda t}P^t u\,dt\}/h$$

$$= \{\int_h^\infty e^{-\lambda(t-h)}P^t u\,dt - \int_0^\infty e^{-\lambda t}P^t u\,dt\}/h$$

$$= -\frac{1}{h}\int_0^h e^{-\lambda t}P^t u\,dt + \int_h^\infty \frac{e^{-\lambda(t-h)} - e^{-\lambda t}}{h}P^t u\,dt$$

$$\longrightarrow -u + \int_0^\infty \lambda e^{-\lambda t}P^t u\,dt = -u + \lambda R_\lambda u\ .$$

Therefore $\mathcal{R}(R_\lambda) \subset \mathcal{D}(A)$ and $AR_\lambda = -1 + \lambda R_\lambda$ ; that is, $(\lambda - A)R_\lambda = 1$ . Also, if $u$ is in $\mathcal{D}(A)$ then, as is easily seen, $P^t u$ is in $\mathcal{D}(A)$ and $AP^t u = P^t A u$ , so that

$$R_\lambda A u = \int_0^\infty e^{-\lambda t}P^t A u\,dt$$

$$= \int_0^\infty e^{-\lambda t}AP^t u\,dt = A\int_0^\infty e^{-\lambda t}P^t u\,dt = AR_\lambda u\ .$$

Thus $R_\lambda(\lambda - A)u = u$ for $u$ in $\mathcal{D}(A)$ . Together with the fact that $(\lambda - A)R_\lambda = 1$ , this means that

$$R_\lambda = (\lambda - A)^{-1}\ , \qquad \operatorname{Re}\lambda > 0\ .$$

Now

$$\lim_{\lambda\to\infty} \lambda R_\lambda u = \lim_{\lambda\to\infty} \lambda\int_0^\infty e^{-\lambda t}P^t u\,dt = u\ , \qquad u \in \mathfrak{X}\ ,$$

and each $\lambda R_\lambda u$ is in $\mathcal{D}(A)$. Thus $A$ is densely defined. Since $(\lambda - A)^{-1}$ is in $L(\mathcal{X})$, $(\lambda - A)^{-1}$ is closed, so $\lambda - A$ is closed, and so $A$ is closed. QED

The Hille-Yosida theorem [26] asserts that if $A$ is a closed, densely defined operator on a Banach space such that $\lambda$ is in the resolvent set of $A$ for $\lambda > 0$ with $\|(\lambda - A)^{-1}\| \leq 1/\lambda$ then $A$ is the infinitesimal generator of a unique contraction semigroup.

Let $A$ be an operator on the Banach space $\mathcal{X}$. A <u>core</u> of $A$ is a linear subspace $\mathcal{D}$ of $\mathcal{D}(A)$ such that $A$ and the restriction of $A$ to $\mathcal{D}$ have the same closure: $\bar{A} = \overline{A|\mathcal{D}}$. For example, if $A$ is a self-adjoint operator on a Hilbert space, a core of $A$ is any linear subspace $\mathcal{D}$ of $\mathcal{D}(A)$ such that the restriction of $A$ to $\mathcal{D}$ is essentially self-adjoint (for if one self-adjoint operator is contained in another, they are equal).

<u>Theorem</u> 2. <u>Let</u> $A_n$, <u>for</u> $n = 1,2,3,\ldots$, <u>and</u> $A$ <u>be the infinitesimal generators of the contraction semigroups</u> $P_n^t$ <u>and</u> $P^t$. <u>Let</u> $\mathcal{D}$ <u>be a core of</u> $A$, <u>and suppose that for all</u> $u$ <u>in</u> $\mathcal{D}$, $u$ <u>is in</u> $\mathcal{D}(A_n)$ <u>for</u> $n$ <u>sufficiently large and</u>

$$A_n u \longrightarrow Au. \tag{3}$$

<u>Then for all</u> $u$ <u>in</u> $\mathcal{X}$,

$$P_n^t u \longrightarrow P^t u \tag{4}$$

<u>uniformly for</u> $t$ <u>in any compact subset of</u> $[0,\infty)$.

<u>Proof.</u> Let $\operatorname{Re}\lambda > 0$. We claim that for all $u$ in $\mathcal{X}$,

$$(\lambda - A_n)^{-1} u \longrightarrow (\lambda - A)^{-1} u. \tag{5}$$

Since the $(\lambda - A_n)^{-1}$ are bounded in norm uniformly in $n$ (by $1/\text{Re}\,\lambda$), we need only show that (5) holds for $u$ in a dense set. Now $(\lambda - A)\mathcal{D}$ is dense because $\mathcal{D}$ is a core and $(\lambda - A)\,\mathcal{D}(A) = \mathcal{X}$. Therefore we may assume that $u = (\lambda - A)v$ with $v$ in $\mathcal{D}$. Then

$$\begin{aligned}\|(\lambda - A_n)^{-1}u - (\lambda - A)^{-1}u\| &= \|(\lambda - A_n)^{-1}(\lambda - A)v - v\| \\ &= \|(\lambda - A_n)^{-1}(\lambda - A_n)v + (\lambda - A_n)^{-1}(A_n - A)v - v\| \\ &= \|(\lambda - A_n)^{-1}(A_n - A)v\| \le \frac{1}{\text{Re}\,\lambda}\|(A_n - A)v\| ,\end{aligned}$$

and this tends to $0$ by (3). Thus (5) holds.

Let $u$ be in $\mathcal{D}$ and let

$$\varphi_n(t) = \begin{cases}(P_n^t u - P^t u)e^{-t}, & t \ge 0 \\ 0, & t < 0 .\end{cases}$$

Since $A$ is densely defined and $\mathcal{D}$ is a core, $\mathcal{D}$ is dense. Therefore if we show that $\varphi_n(t)$ converges uniformly in $t$ to $0$, we are through, since the $P_n^t$ are bounded in norm uniformly in $n$ (by 1). Now

$$\frac{d}{dt}\varphi_n(t) = (P_n^t A_n u - P^t A u)e^{-t} - (P_n^t u - P^t u)e^{-t}$$

for $t \ge 0$, and this is bounded in norm uniformly in $n$ and $t$, by (3). Thus the $\varphi_n(t)$ are equi-uniformly continuous. Therefore, in order to show that $\varphi_n(t) \longrightarrow 0$ uniformly in $t$ we need only show that $(\varphi_n * \rho)(t) \longrightarrow 0$ uniformly in $t$, for all $C^\infty$ functions $\rho\colon \mathbb{R} \longrightarrow \mathbb{R}$ with compact support, where $\varphi_n * \rho$ is the convolution

$$(\varphi_n * \rho)(t) = \int_{-\infty}^{\infty} \varphi_n(t-s)\rho(s)\,ds .$$

The Fourier transform of $\varphi_n * \rho$ is $\hat{\varphi}_n \hat{\rho}$, where

$$\hat{\varphi}_n(\lambda) = \int_{-\infty}^{\infty} \varphi_n(t) e^{-i\lambda t} dt = (1 + i\lambda - A_n)^{-1} u - (1 + i\lambda - A)^{-1} u$$

and

$$\hat{\rho}(\lambda) = \int_{-\infty}^{\infty} \rho(t) e^{-i\lambda t} dt$$

is in $L^1(\mathbb{R})$. By the Lebesgue dominated convergence theorem and the Fourier inversion formula,

$$(\varphi_n * \rho)(t) = \frac{1}{2\pi} \int_{-\infty}^{\infty} \hat{\varphi}_n(\lambda) \rho(\lambda) e^{i\lambda t} d\lambda \longrightarrow 0$$

uniformly in $t$. QED

<u>Theorem</u> 3. <u>Let</u> $T$ <u>be in</u> $L(\mathfrak{X})$ <u>with</u> $\|T\| \leq 1$. <u>Then</u>

$$t \rightsquigarrow e^{t(T-1)}$$

<u>is a contraction semigroup. For all</u> $u$ <u>in</u> $\mathfrak{X}$,

(6) $$\|(e^{n(T-1)} - T^n)u\| \leq \sqrt{n}\, \|(T-1)u\| .$$

<u>Proof.</u> For $t \geq 0$,

$$\|e^{t(T-1)}\| = \|e^{-t} \sum_{k=0}^{\infty} \frac{t^k T^k}{k!}\| \leq e^{-t} \sum_{k=0}^{\infty} \frac{t^k}{k!} = 1 .$$

The function $t \rightsquigarrow e^{t(T-1)}$ is continuous from $[0,\infty]$ to $L(\mathfrak{X})$ and a fortiori it is strongly continuous, so it is a contraction semigroup.

For any $u$ in $\mathfrak{X}$,

$$\|(e^{n(T-1)} - T^n)u\|$$

$$= \|e^{-n} \sum_{k=0}^{\infty} \frac{n^k}{k!} (T^k - T^n)u\| \le e^{-n} \sum_{k=0}^{\infty} \frac{n^k}{k!} \|(T^k - T^n)u\|$$

$$\le e^{-n} \sum_{k=0}^{\infty} \frac{n^k}{k!} \|(T^{|k-n|} - 1)u\| \le e^{-n} \sum_{k=0}^{\infty} \frac{n^k}{k!} |k-n| \|(T-1)u\| .$$

By the Schwarz inequality, applied to the sequences $|k-n|$ and 1 with the weights $n^k/k!$ ,

$$e^{-n} \sum_{k=0}^{\infty} \frac{n^k}{k!} |k-n| \le \sqrt{\varphi(n)}$$

where

$$\varphi(n) = e^{-n} \sum_{k=0}^{\infty} \frac{n^k}{k!} (k-n)^2 .$$

We see that

$$\frac{d\varphi(n)}{dn} = 1$$

and $\varphi(0) = 0$ , so that $\varphi(n) = n$ . This proves (6). QED

<u>Theorem</u> 4 (<u>Chernoff's theorem</u>). <u>Let</u> $F: [0,\infty) \to L(\mathcal{X})$ <u>with</u> $\|F(t)\| \le 1$ <u>for all</u> $t$ <u>in</u> $[0,\infty)$ <u>and</u> $F(0) = 1$ . <u>Let</u> $P^t$ <u>be a contraction semigroup on</u> $\mathcal{X}$ <u>with infinitesimal generator</u> $A$ , <u>and let</u> $\mathcal{D}$ <u>be a core of</u> $A$ . <u>Suppose that</u>

$$\lim_{h \to 0} \frac{F(h)-1}{h} u = Au , \qquad u \in \mathcal{D} .$$

<u>Then for all</u> $u$ <u>in</u> $\mathcal{X}$ ,

(7) $$\lim_{n \to \infty} F(\tfrac{t}{n})^n u = P^t u$$

<u>uniformly for</u> $t$ <u>in any compact subset of</u> $[0,\infty)$ .

Proof. Fix $t > 0$ and let

$$C_n = \frac{n}{t}(F(\tfrac{t}{n}) - 1) .$$

By Theorem 3, $\frac{t}{n} C_n$ is the infinitesimal generator of a contraction semigroup, and since $t/n \geq 0$, $C_n$ is itself the infinitesimal generator of the contraction semigroup

$$t \rightsquigarrow e^{tC_n} .$$

By Theorem 2, for all $u$ in $\mathcal{X}$

$$e^{tC_n}u \longrightarrow P^t u$$

uniformly for $t$ in any compact subset of $[0,\infty)$. But by Theorem 3,

$$\|(e^{tC_n} - F(\tfrac{t}{n})^n)u\| \leq \sqrt{n}\|(F(\tfrac{t}{n}) - 1)u\| \leq \frac{t}{\sqrt{n}}\left\| \frac{F(\frac{t}{n}) - 1}{(\frac{t}{n})} u \right\| ,$$

which converges to 0 (uniformly for $t$ in any compact subset of $[0,\infty)$) for $u$ in $\mathcal{D}$. Thus (7) holds for a dense set of $u$'s and hence for all $u$ in $\mathcal{X}$, since the $F(\frac{t}{n})^n$ are bounded in norm by 1. QED

Theorem 5 (Trotter product formula). Let

$$A, B, \quad \text{and} \quad \overline{A + B}$$

be the infinitesimal generators of the contraction semigroups

$$P^t, Q^t, \quad \text{and} \quad R^t$$

on the Banach space $\mathcal{X}$. Then for all $u$ in $\mathcal{X}$,

$$(8) \qquad R^t u = \lim_{n \to \infty} \left(P^{\frac{t}{n}} Q^{\frac{t}{n}}\right)^n u ,$$

uniformly for $t$ in any compact subset of $[0,\infty)$.

Proof. $\mathcal{D}(A+B) = \mathcal{D}(A) \cap \mathcal{D}(B)$ is a core for $\overline{A+B}$. Let $F(t) = P^t Q^t$. For $u$ in $\mathcal{D}(A+B)$,

$$F(t)u = P^t Q^t u = P^t(u + tBu + o(t)) = u + tAu + tBu + o(t) .$$

Therefore (8) holds by Theorem 4. QED

We state a special case of this explicitly.

Theorem 6. Let $A$ and $B$ be self-adjoint operators on a Hilbert space $\mathcal{H}$, and suppose that $A+B$ is essentially self-adjoint. Then for all $u$ in $\mathcal{H}$,

$$e^{it(\overline{A+B})}u = \lim_{n \to \infty} \left(e^{i\frac{t}{n}A} e^{i\frac{t}{n}B}\right)^n u ,$$

uniformly for $t$ in any compact subset of $(-\infty,\infty)$.

An operator $A$ on a Hilbert space is called skew-adjoint in case $iA$ is self-adjoint; that is, in case $A = -A^*$. It is called essentially skew-adjoint in case $iA$ is essentially self-adjoint; that is, in case $\bar{A} = -A^*$. If $A$ is skew-adjoint then $A$ and $-A$ are infinitesimal generators of contraction semigroups $e^{tA}$ and $e^{-tA}$ which together make up the strongly continuous one-parameter unitary group $e^{tA}$ for $-\infty < t < \infty$.

If $A$ and $B$ are two operators, then

$$[A,B] = AB - BA$$

is called their Lie product or commutator. Notice that if $A$ and $B$ are in $L(\mathcal{H})$ and are skew-adjoint so is their Lie product.

Theorem 7. *Let* $A$ *and* $B$ *be skew-adjoint operators on a Hilbert space* $\mathcal{H}$, *and suppose that the restriction of* $[A,B]$ *to* $\mathcal{D} = \mathcal{D}(AB) \cap \mathcal{D}(BA) \cap \mathcal{D}(A^2) \cap \mathcal{D}(B^2)$ *is essentially skew-adjoint. Then for all* $u$ *in* $\mathcal{H}$,

$$e^{t\overline{[A,B]}}u = \lim_{n\to\infty}\left(e^{-\sqrt{\frac{t}{n}}A}\,e^{-\sqrt{\frac{t}{n}}B}\,e^{\sqrt{\frac{t}{n}}A}\,e^{\sqrt{\frac{t}{n}}B}\right)^n u, \tag{9}$$

*uniformly for* $t$ *in any compact subset of* $[0,\infty)$.

Proof. Since $[A,B]|\mathcal{D}$ is essentially skew-adjoint, $[A,B]$, which has domain $\mathcal{D}(AB) \cap \mathcal{D}(BA)$, is essentially skew-adjoint, and they have the same closure $\overline{[A,B]}$, which has $\mathcal{D}$ as a core. Let

$$F(t) = e^{-tA}e^{-tB}e^{tA}e^{tB}.$$

For $u$ in $\mathcal{D}$,

$$\begin{aligned} F(t)u &= e^{-tA}e^{-tB}e^{tA}(1 + tB + \tfrac{t^2}{2}B^2)u + o(t^2) \\ &= e^{-tA}e^{-tB}(1 + tA + \tfrac{t^2}{2}A^2 + tB + t^2AB + \tfrac{t^2}{2}B^2)u + o(t^2) \\ &= e^{-tA}(1 - tB + \tfrac{t^2}{2}B^2 + tA - t^2BA + \tfrac{t^2}{2}A^2 + tB - t^2B^2 + t^2AB + \tfrac{t^2}{2}B^2)u + o(t^2) \\ &= (1 - tA + \tfrac{t^2}{2}A^2 - tB + t^2AB + \tfrac{t^2}{2}B^2 + tA - t^2A^2 - t^2BA + \tfrac{t^2}{2}A^2 + tB - t^2AB \\ &\qquad - t^2B^2 + t^2AB + \tfrac{t^2}{2}B^2)u + o(t^2) \\ &= (1 + t^2[A,B])u + o(t^2). \end{aligned}$$

By Chernoff's theorem, (9) holds. QED

Notes and references

The parallel between quantum mechanics and classical mechanics is much closer if one considers only the Hamiltonian formulation of classical mechanics. We shall not give this formulation here, as this chapter is devoted to kinematics only. See [1], [2], and [3].

§1. As general refrences see [4], [5], and [6].

§2. See [4], [5], [7], [8].

§3. See [9], [10], [11], and [7, Chapter IX]. A by-product of our proof of the Sternberg linearization theorem was a proof of the existence of the local stable and unstable manifolds. This can be proved directly for any elementary critical point without restrictive smoothness assumptions, see [7].

For a discussion of problems relating to the local structure of Hamiltonian vector fields in the neighborhood of a critical point, see [1] and [11].

Linearization of analytic vector fields is studied in [29].

§4. See [13, pp.30-36].

§5. For accounts of Hilbert space, see [14] and [15]. The first two chapters of [16] have an account of bounded operators on Hilbert space.

Another approach to the spectral theorem, in some ways preferable to the one we gave, is the following. First one proves Stone's theorem, perhaps deducing it as a special case of the Hille-Yosida theorem concerning contraction semigroups on a Banach space. Then given a self-adjoint operator A one has the strongly continous one-parameter unitary group $U(t)$ with infinitesimal generator $iA$. If $u$ is in the Hilbert space, then $t \rightsquigarrow (u,U(t)u)$ is a continuous function of positive type, and so by Bochner's theorem there is a unique measure $\mu = \mu_u$ such that

$$(u,U(t)u) = \int_{-\infty}^{\infty} e^{it\lambda} d\mu(\lambda) .$$

Let $\mathcal{H}_u$ be the smallest closed linear subspace containing $u$ and invariant under $U(t)$. Then $A$ on $\mathcal{H}_u$ is unitarily equivalent to multiplication by the identity function $\lambda \rightsquigarrow \lambda$ on $L^2(\mathbb{R},\mu)$.

A reference for measure theory is [17]. If $X$ is a locally compact Hausdorff space, we always use the term "measure" to be synonymous with Radon measure; that is, a regular Borel measure which is finite on all compact sets.

§6. There is an account of commutative multiplicity theory in [16, Chapter III]. For σ-functions, see [18]. Theorem 1 is usually proved by introducing the maximal ideal space, see [20].

§7. See [14], [15], [21], [22].

§8. Trotter [24] shows that the strong convergence of $(\lambda-A_n)^{-1}$ to $(\lambda-A)^{-1}$ implies the strong convergence of $P_n^t$ to $P^t$. Chernoff [23] by-passes this difficult result, but the main point of his approach is the use he makes of the estimate (6) of Theorem 3.

* * *

*We have discussed vector fields and their flows only locally.* For the notions of a differentiable manifold and the flow generated by a vector field, see [2], [5], [6], or [27]. We shall use the term "manifold" to mean a finite dimensional, Hausdorff, second countable differentiable manifold. The Hausdorff property ensures that the flow generated by a vector field on a manifold is unique. However, the flow is in general only locally defined, as the orbit of a point may run off the manifold (if it is not compact) at a finite time. A vector field $X$ on a manifold $M$ which generates a one-parameter group of diffeomorphisms of $M$ is called complete. The vector fields of interest in dynamics are almost never complete. For example, in the two body problem with Newtonian gravitational attraction, if the angular momentum is zero, the two bodies will collide with infinite velocities at some finite time. However, this happens only for a set of initial conditions in phase space $M$ of measure $0$. For a finite dimensional manifold $M$, the notion of a set of measure $0$ has an invariant meaning. We say that a vector field $X$ on $M$ is almost complete in case for each $t \geq 0$ there is a closed set $E_t$ of measure $0$, diffeomorphisms $U(s)$ from $M-E_t$ to an open subset of $M$ for $|s| \leq t$ such that $U(s_1)U(s_2)x = U(s_1+s_2)x$ for $x$ in $M-E_t$ and $|s_1|, |s_2|, |s_1+s_2| < t$ and such that for each $x$ in $M-E_t$, $U(s)x$ is tangent to $Xx$ at $s = 0$.

Stone's theorem is an analogue of the existence and uniqueness theorem for flows generated by a vector field. If $X$ is a complete or almost complete vector field on the manifold $M$ which has a smooth measure $\mu$ invariant under the corresponding flow $U(t)$, then $f \rightsquigarrow U(t)f$ where $(U(t)f)(x) = f(U(-t)x)$ is a strongly continuous one-parameter unitary group on $L^2(M,\mu)$. More generally, one may form the Hilbert space $\mathcal{H}_0(M)$ of all $\sigma$-functions $f\sqrt{d\mu}$ such that in each local coordinate system we may choose $\mu$ to be smooth (see [19]). This is an intrinsic notion, so diffeomorphisms of M induce unitary operators on $\mathcal{H}_0(M)$. Vector fields also act on $\mathcal{H}_0(M)$, with the domain of all smooth $\sigma$-functions, (those for which we may choose both f and $\mu$ to be smooth). Consequently the enveloping algebra of the vector fields, which is the algebra of partial differential operators, acts on $\mathcal{H}_0(M)$. This should be a rewarding subject for investigation.

Commutative multiplicity theory is rather tedious, but it does accomplish a complete classification of self-adjoint operators. The classification of vector fields is much more difficult. The Sternberg linearization theorem classifies the generic vector field locally, but it leaves out the most interesting case, that of Hamiltonian vector fields. Recently there has been a lot of attention devoted to the investigation of generic global properties of vector fields on manifolds, see [28] and [2].

I do not know of any analogue of the Friedrichs extension theorem in classical Hamiltonian mechanics.

Let M be a differentiable manifold (phase space), x a point of M (state of the system), f a real function on M (dynamical variable). Then the value of the dynamical variable f, given that the state of the system is x, is f(x).

Let $\mathcal{H}$ be a Hilbert space, u a unit vector (state of the quantum mechanical system), A a self-adjoint operator on $\mathcal{H}$ (dynamical variable). Then the value of the dynamical variable A, if an observation is made to determine its value, may be any number in the spectrum of A , and the probability that it will lie in a Borel set $B \subset \mathbb{R}$ , given that the state of the system is u , is

$$\langle u, E_B u \rangle ,$$

where $E_B$ is the spectral projection $\chi_B(A)$ .

Let U(t) be a one-parameter group of diffeomorphisms of M , or of unitary operators on $\mathcal{H}$ . We may consider the action of U(t) on M (or $\mathcal{H}$ ) , keeping the dynamical variables fixed. This is called the Schrödinger picture. Or we may keep the state of the system fixed and let U(t) act on the dynamical variables via $f \rightsquigarrow U(t)f$ (or $A \rightsquigarrow U(-t)AU(t)$). This is called the Heisenberg picture.

[1]. Jürgen Moser, Lectures on Hamiltonian Systems, and W. T. Kyner, Rigorous and Formal Stability of Orbits about an Oblate Planet, Memoirs of the American Mathematical Society Number 81, 1968.

[2]. Ralph Abraham, with the assistance of Jerrold E. Marsden, Foundations of Mechanics: A mathematical exposition of classical mechanics with an introduction to the qualitative theory of dynamical systems and applications to the three-body problem, W. A. Benjamin, 1967.

[3]. Herbert Goldstein, Classical Mechanics, Addison-Wesley, 1950.

[4]. J. Dieudonné, Foundations of Modern Analysis, Academic Press, 1960.

[5]. Serge Lang, Introduction to Differentiable Manifolds, Interscience, 1962.

[6]. Ralph Abraham and Joel Robbin, Transversal Mappings and Flows, W. A. Benjamin, 1967.

[7]. Philip Hartman, Ordinary Differential Equations, John Wiley & Sons, 1964.

[8]. Earl A. Coddington and Norman Levinson, Theory of Ordinary Differential Equations, McGraw-Hill, 1955.

[9]. Shlomo Sternberg, Local contractions and a theorem of Poincaré, Amer. J. of Math. 79(1957), 809-824.

[10]. -----, On the structure of local homeomorphisms of Euclidean n-space, II, ibid. 80(1958), 623-631.

[11]. -----, The structure of local homeomorphisms, III, ibid. 81(1959), 578-604.

[12]. Joel W. Robbin, On the existence theorem for differential equations, Proc. Amer. Math. Soc. 19(1968), 1005-1006.

[13]. Edward Nelson, Tensor Analysis, Mathematical Notes, Princeton University Press, 1967.

[14]. Marshall Harvey Stone, Linear Transformations in Hilbert Space and their Applications to Analysis, American Mathematical Society Colloquium Publications, Vol. XV, 1932.

[15]. Frigyes Riesz and Béla Sz.-Nagy, Functional Analysis, Translated by Leo F. Boron, Frederick Ungar, 1955.

[16]. Paul R. Halmos, Introduction to Hilbert Space and the Theory of Spectral Multiplicity, Chelsea, 1951.

[17]. -----, Measure Theory, D. Van Nostrand, 1950.

[18]. Laurent Schwartz, Généralisation des espaces $L^p$, Publ. Inst. Statist. Univ. Paris 6(1957), 241-250.

[19]. George W. Mackey, Mathematical Foundations of Quantum Mechanics, W. A. Benjamin, 1963.

[20]. Lynn H. Loomis, An Introduction to Abstract Harmonic Analysis, D. Van Nostrand, 1953.

[21]. K. O. Friedrichs, Spektraltheorie halbbeschrankter Operatoren, Math. Ann. 109(1934), 465-487, 685-713.

[22]. J. L. Lions, Equations differentielles operationnelles et problèmes aux limites, Springer-Verlag, 1961.

[23]. Paul Chernoff, Note on product formulas for operator semi-groups, J. Functional Analysis 2(1968), 238-242.

[24]. H. F. Trotter, Approximation of semigroups of operators, Pacific J. Math. 8(1958), 887-919.

[25]. -----, On the product of semigroups of operators, Proc. Am. Math. Soc. 10(1959), 545-551.

[26]. K. Yosida, On the differentiability and the representation of one-parameter semigroups of linear operators, J. Math. Soc. Japan 1(1948), 15-21.

[27]. Shlomo Sternberg, Lectures on Differential Geometry, Prentice-Hall, 1965.

[28]. S. Smale, Differentiable dynamical systems, Bull. Amer. Math. Soc. 73(1967), 747-817.

[29]. Carl Ludwig Siegel, Über die Normalform analytischer Differentialgleichungen in der Nähe einer Gleichgewichtslösung, Nachrichten der Akademie der Wissenschaften Göttingen, Math.-Phys. Kl Math.-Phys.-Chem. Abt. 1952(1952), 21-30.